GCSE
Geography
for WJEC A

Revision Guide

Dirk Sykes & Stacey Burton-McCabe

HODDER
EDUCATION
AN HACHETTE UK COMPANY

The Publishers would like to thank the following for permission to reproduce copyright material:

Photo credits p.11 © Stuart Currie; **p.19** © www.live-the-solution.com, used with kind permission; **p.27** © Julien Grondin/iStockphoto.com; **p.34** © Royalty-Free/Corbis; **p.49** © Jon Spaull/WaterAid, used with kind permission; **p.53** © Andy Owen; **p.64** © Debbie Allen; **p.77** © Dirk Sykes; **p.79** © Dirk Sykes; **p.80** © Ian Dagnall/Alamy.

Maps on pages **57** and **87** reproduced from Ordnance Survey mapping with the permission of the Controller of HMSO, © Crown copyright. All rights reserved. Licence no. 1000364700.

Text acknowledgements p.46 Adapted from the Millennium Development Goals report 2009, http://www.un.org/ millenniumgoals/ © Copyright United Nations Development Programme, 2006. All Rights Reserved; **p.48** Adapted from the Millennium Development Goals report 2009, http://www.un.org/millenniumgoals/ © Copyright United Nations Development Programme, 2006. All Rights Reserved. **p.49** Article and use of logo taken from http://www.oxfam.org/en/development/ghana/hygiene-education © Oxfam International; **p.56** Reproduced by kind permission of the BBC and taken from the BBC news website: http://news.bbc.co.uk/1/hi/wales/6356073.stm; **p.80** Mission statement © Cardiff Bay Development Corporation (CBDC); **p.84** Reproduced by kind permission of the BBC and taken from the BBC news website: http://news.bbc.co.uk/1/hi/wales/7349546.stm **p.91** Diagram adapted from www.geographyfieldwork.com and © Barcelona Field Studies Centre.

Every effort has been made to trace all copyright holders, but if any have been inadvertently overlooked the Publishers will be pleased to make the necessary arrangements at the first opportunity.

Although every effort has been made to ensure that website addresses are correct at time of going to press, Hodder Education cannot be held responsible for the content of any website mentioned in this book. It is sometimes possible to find a relocated web page by typing in the address of the home page for a website in the URL window of your browser.

Hachette UK's policy is to use papers that are natural, renewable and recyclable products and made from wood grown in sustainable forests. The logging and manufacturing processes are expected to conform to the environmental regulations of the country of origin.

Orders: please contact Bookpoint Ltd, 130 Milton Park, Abingdon, Oxon OX14 4SB. Telephone: (44) 01235 827720. Fax: (44) 01235 400454. Lines are open 9.00–5.00, Monday to Saturday, with a 24-hour message answering service. Visit our website at www.hoddereducation.co.uk

© Dirk Sykes and Stacey Burton-McCabe 2010
First published in 2010 by
Hodder Education,
An Hachette UK Company
338 Euston Road
London NW1 3BH

Impression number   8 7 6
Year   2015 2014 2013

Cover photo: G.A.P Adventures' M/S Explorer cruise ship moored near a colony of King penguins (*Aptenodytes patagonicus*) on South Georgia Island, © Martin Harvey/Corbis.
Illustrations by Tim Oliver, Barking Dog Art, Oxford Designers and Illustrators and DC Graphic Design Ltd.
Editorial by Hart McLeod, Cambridge
Typeset in 10.5pt Trade Gothic by Hart McLeod, Cambridge
Printed in India

A catalogue record for this title is available from the British Library

ISBN: 978 1444 100914

Contents

Introduction   4

Unit 1 The Core
A The Physical World

Theme 1 – Water	8
Theme 2 – Climate Change	15
Theme 3 – Living in an Active Zone	21

B A Global World

Theme 4 – Changing Populations	29
Theme 5 – Globalisation	37
Theme 6 – Development	43

Unit 2 Option Topics
A The Physical World

Theme 7 – Our Changing Coastline	51
Theme 8 – Weather and Climate	58
Theme 9 – Living Things	63

B A Global World

Theme 10 – Tourism	71
Theme 11 – Retail and Urban Change	79
Theme 12 – Economic Change and Wales	88

Introduction

Why should I use this guide?

This revision guide accompanies the GCSE Geography for WJEC Specification A course and is designed to help you get the best possible results in your examinations.

This revision guide is different from the textbooks you may have used for your course. *GCSE Geography for WJEC A Core* and *Option Topics*, for example, give you lots of information, activities and case studies. The aim of this revision guide is to help you gain examination success.

This guide gives you essential information which should remind you of what you have covered in your course. Activities are included to test your understanding of a topic and to help you revise. You will also find advice on study/revision skills from an expert in the field, so that you can make the most effective use of your time. The guide also gives you inside information from an experienced examiner, which will help you play the examination game and maximise your grade. There are many candidates who are excellent geographers but fail to obtain the grade that reflects their ability because they do not understand what the examiner is looking for in an answer. This book shows you exactly what the examiner is after.

How should I use this guide?

This revision guide contains a number of repeated features. These are intended to help you *actively* work through your revision schedule as painlessly as possible! They are highlighted by a number of symbols, each representing a different feature.

The Essentials

Here you are given the bare bones of knowledge for a given topic. You must use this knowledge as the skeleton on which you add detail. Your own notes and the textbook accompanying Specification A will provide this detail. Specific knowledge is needed to gain full marks in examination questions.

Go Active

These activities/tasks have been designed to test your understanding and focus your revision. If you can complete all of these activities first time round, congratulate yourself and give yourself a reward. If you find some of the activities challenging, read over your notes again and speak to your teacher and your friends. It is worth remembering that your peers can often be a fabulous source of information – you may even decide to study as a group with friends. Find the best way of revising for you.

Inside Information

These sections give examples of examination questions and actual candidate answers. They offer an insight into exactly what the examiner wants. They also analyse mark schemes so that you can become your own examiner and improve your grade.

Exam Spotlight

These sections will guide you through exam preparation and help you formulate strategies for getting the best out of the time you have in the examination room. You will also be given plenty of opportunities to practise exam-style questions, as well as advice on how to answer them.

Case Study

Here you will be given punchy case studies to support some of the themes you have studied. Examiners are impressed when you can give real examples in your answers. Case studies are often needed to take an answer into Level 3 of a mark scheme.

Getting to know the specification

The WJEC Geography A Specification

The content of the specification is divided into two units: 'Core' and 'Option Topics'.
Both units are divided into three physical and three human themes.

Unit 1 – Core

You will cover all six topics in the physical and human themes.

Unit 1 – Core			
A The Physical World	**1 Water** River processes and landforms Managing rivers	**2 Climate Change** Causes and effects Reducing its impact	**3 Living in an Active Zone** Hazards at plate margins Reducing the risk
B A Global World	**4 Changing Populations** World population distribution Future changes in distribution and structure	**5 Globalisation** Trends in globalisation Impacts of globalisation	**6 Development** Measuring patterns of development Achieving the Millennium Development Goals

Unit 2 – Options

Your teacher will choose three themes – one physical, one human and one other from the options below.

Unit 2 – Options			
A Physical Options	**7 Our Changing Coastline** Coastal processes and landforms Managing coasts Future coastlines	**8 Weather and Climate** Climate patterns in the UK Weather hazards Reducing the risks	**9 Living Things** The living planet Management Alternative futures
B Human Options	**10 Tourism** The changing nature of tourism The impact of tourism Sustainable growth of tourism	**11 Retail and Urban Change** The changing city centre Changing patterns of retailing Alternative futures	**12 Economic Change and Wales** Current patterns of work and employment Future employment Future for energy in Wales

Assessment through WJEC Specification A

Students can be entered for the Foundation Tier (grades C–G) or the Higher Tier (grades A*–D). It is possible to combine Foundation and Higher Tier entry to give you your final grade. The division of the specification into units means that you can take Unit 1 in your first year of study and Units 2 and 3 at the end of the course. Alternatively, all units may be taken at the end of the course if preferred.

Unit 1: Core (40%)	This paper is 1 hour 45 minutes long and consists of six compulsory structured data response questions; one question from each of the core themes.
Unit 2: Options (35%)	This paper is 1 hour 15 minutes long and consists of three in-depth data response questions, with extended writing. Students must answer one question from the physical options, one from the human options and one other.
Unit 3: Geographical Enquiry (25%)	This controlled assessment consists of an enquiry based on fieldwork (10%) and a problem-solving, decision-making exercise (15%).

Preparing for the exam

Top tips for exam success (the rules of the game)!

Remember that every question contains a balance between testing knowledge, understanding and skills.

During the examination think about the following ten basic rules:

1 Read and follow instructions carefully.
2 Study any resources carefully and use them effectively in your answer.
3 Understand the meaning of command words.
4 Identify key terms and use them to plan your answer.
5 Answer in sufficient detail and depth – be guided by the marks available for the question.
6 Develop points made – turn one mark into two.
7 Plan answers that require extended writing.
8 Use case studies to exemplify your answers.
9 Use key geographical terms.
10 Draw good sketch maps and diagrams where they will help your answer.

Go Active

Do you understand command words? All examination questions contain command words. In order to gain full marks you need to understand what these words mean; what do they want you to do? Match the command words below with their meanings.

When you have finished, check your answers with the table on page 96.

A	Study		I	Spot and single out
B	Locate		2	Add explanatory notes to a map or diagram
C	Outline		3	A brief summary
D	Describe		4	Tell the examiner what you see/use lots of adjectives
E	Explain		5	Say how it is similar and different
F	Identify		6	Look carefully at
G	Suggest		7	Put forward an idea
H	Give reasons		8	Identify and mark
I	Compare		9	Say why or account for
J	Label		10	Means the same as explain
K	Annotate		11	Write down where a feature or place is

Use this table to record your answers. One has been done for you.

A	B	C	D	E	F	G	H	I	J	K
6										

Effective revision

There is a huge variety of different techniques that people use to revise. You need to identify those which work for you. Consider the following questions:

- Do you need complete silence to revise?
- Does music help you concentrate?
- When do you like to revise? Morning, evening, after lunch?
- For how long can you concentrate without taking a break? There is a common belief that the average person struggles to concentrate for longer than 20 minutes without a change in focus.
- Do you have a room where you can revise? Is it well lit with natural light?
- What parts of your social life are you prepared to give up for revision?
- How many weeks are you going to devote to revision? How are you going to divide those weeks, e.g. a whole day per subject, a different subject in the morning, afternoon and evening?
- Do you work well as part of a study group?
- Does your school provide revision classes?

Once you've decided what conditions you need for effective revision, you need to construct your revision timetable. Make sure you start early enough to cover the material and remember to build in rewards. It is important to balance work with leisure. Remember also that you are not alone. Your teacher, parents and friends can all help. If at any time you feel unable to cope, be sure to talk to someone about it.

Active revision

Your revision could be active or passive. Passive revision involves simply reading your notes. It is something which can only be used in short bursts, perhaps in the hours before an exam, and only stores information in your short term memory. After a short period of passive revision your attention will begin to wander – perhaps you will find yourself staring at a poster on the wall or listening to music.

Active revision involves you doing something. This is more likely to hold your attention, which will help you remember facts and information more effectively. It may even help you produce something that you can use later in the revision process.

A number of active revision activities are suggested in this revision guide, although you may decide to devise your own. You may wish to use a revision exercise book, in which you can produce your own neat and punchy revision notes using your class work and textbook.

Other techniques that may work for you include:

- revision cards for important case studies. Remember to use colours when highlighting key words and to make a note of causes, effects and solutions.
- mind maps to link key ideas in each theme. Remember mind maps need pictures. Drawing simple pictures is an extremely useful way of helping your memory retain information.
- a wheel of knowledge for each theme to identify your strengths and weaknesses.
- cards with questions to test your knowledge and understanding. Produce another set of cards with answers to match the questions.
- a book of important diagrams and maps that need to be learnt. Remember to label and annotate each diagram to identify and explain its main features.

The day of the exam

On the day of the examination you should:

- get up early, spend an hour looking over your revision notes (use your short term memory)
- have breakfast
- arrive at the examination room with plenty of time to spare
- listen to the invigilator
- know your centre and candidate number
- have the right equipment for the exam (remember a spare pen)
- make sure you answer all questions
- remember the rules of the game
- never leave the examination room until you have used the complete time available.

Good luck!

What are river processes and what landforms do they create?

What processes are associated with rivers?

The Essentials

Erosion	Transport	Deposition
Erosion is the wearing away of land. Rivers can erode in four ways: • **Hydraulic action** – the sheer force of flowing water washes away any loose material on the bed and banks. • **Abrasion** (also known as **corrasion**) – stones carried in the river are washed into the bed and banks, wearing them away. Sometimes stones get trapped in a dip, where they swirl around and create potholes. • **Solution** (also known as **corrosion**) – the slightly acidic river water dissolves rocks made of calcium carbonate ($CaCO_3$). • **Attrition** – stones collide together and are broken down becoming smaller and rounder.	The material carried by a river is called its load. The load is transported in four ways: • **Traction** – stones are rolled along the river bed by the force of flowing water. • **Saltation** – small stones are bounced along the river bed by flowing water. • **Suspension** – particles of silt and clay float and are carried along in the flowing water. • **Solution** – some minerals dissolve in water. Limestone, for example, slowly dissolves.	If a river slows down because there is less water flowing in it, or the land is less steep, the river will have less energy to carry its load. Some of this material therefore will be dropped (deposited). Rivers slow down and deposit material on the inside of meander bends, in shallow water and when they reach the sea.

The Essentials

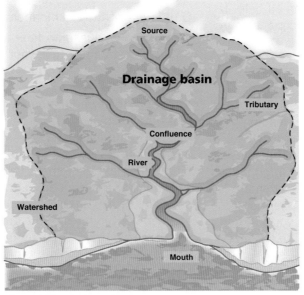

Figure 1 The drainage basin

The drainage basin

- A **drainage basin** is an area of land drained by a river and its tributaries.
- Rivers have their **source** in the mountains.
- Precipitation that falls within the boundary of the **watershed** drains into a river system.
- Smaller **tributary** streams join the main river as it journeys towards the sea. The two will eventually meet at the river's **mouth**.
- During this journey, the river creates landforms and develops the landscape.

What landforms result from these processes?

V-shaped river valleys

- In the mountains, the river cuts into the land as it flows down steep slopes.
- This vertical erosion, together with the movement of weathered material at the sides of the river, creates the classic **V-shaped valley**.
- The river winds around obstacles of hard rock creating **interlocking spurs**. These ridges interlock like the teeth of a zip fastener.

Waterfalls

Eventually the overhang collapses due to lack of support and gravity causes it to fall

Hard resistant rock

As water falls over the lip, more of the rock is eroded by hydraulic action – sheer force of water as splash back against soft rock

Gradually the waterfall retreats upstream leaving a steep-sided gorge

Soft less resistant rock (this is easily eroded)

Hard rock is undercut by erosion of the soft rock

A plunge pool is formed by the force of water hitting soft rock below and deepened by rocks rubbing against the bed (corrasion)

Original position of waterfall

Figure 2 Waterfall development

- A **waterfall** occurs when a layer of hard rock lies over a layer of softer rock.
- The hard rock is less easily eroded, creating a steep slope and **rapids**.
- Eventually a drop develops over which the water falls.

Meanders

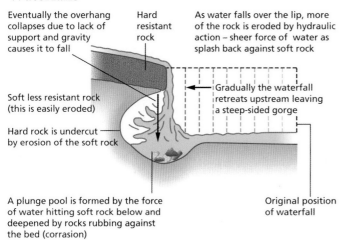

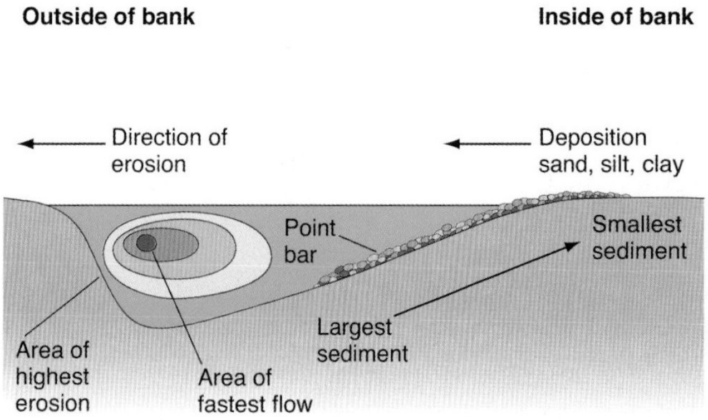

Outside of bank

Inside of bank

Direction of erosion

Deposition sand, silt, clay

Point bar

Smallest sediment

Largest sediment

Area of highest erosion

Area of fastest flow

Figure 3 Features of meanders

- A **meander** is a curve in a river formed by sideways (lateral) erosion.
- Meander bends are found in lower areas of the river's journey to the sea.
- Erosion on the outside of the meander bend, where water is deeper and the river flows faster, and deposition on the inside of the bend, mean that the river channel 'migrates'. This forms a flat valley floor, known as a **floodplain**, over which the river floods in periods of high rainfall.

Oxbow lakes

- As the meander continues to erode sideways, the loop of the bend becomes tighter. If it becomes too tight, the river may simply cut across the neck of the meander to form a straight river channel.
- The loop is cut off from the main channel and forms an oxbow lake.

Floodplains

- As the river nears its mouth it has a large discharge and the river channel is deep and wide. The valley is also wide with an extensive floodplain.
- Each time the river floods, silt is deposited on the flood plain. This creates fertile soils.
- Natural **levees** are raised banks of deposited material found either side of the river channel. Levees are formed because coarse and heavy sediments are deposited first, near the channel, when the river floods.

Deltas

- A **delta** is a feature found at the mouth of a river.
- When a river enters the sea, it loses energy and can no longer carry as much material. Its load is then deposited. If this happens at a faster rate than that at which the sea can remove the material, a delta may form.
- Deltas can provide rich farmland but often suffer from flooding e.g. the Nile and Ganges.

Go Active

1 Choose one landform you would find in a drainage basin and draw a series of diagrams to explain how it is formed.

2 Label the diagrams with appropriate key words.

How do these landforms and processes affect the lives of people living along rivers?

Go Active

There are many ways in which people use rivers and in which rivers affect the lives of people.

1 How many ways can you think of in which rivers and people affect each other? Record your thoughts in a spider diagram, such as that shown. You can add more circles if needed.

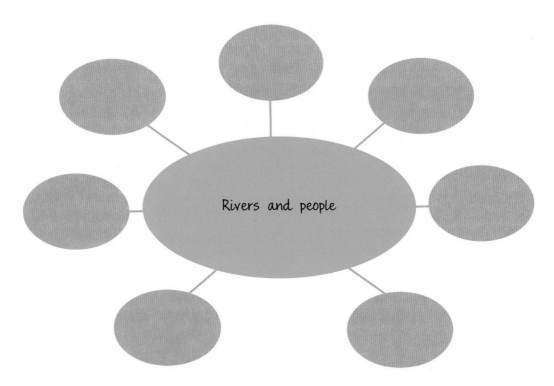

Rivers and people

Example 1 – Tourism

Case Study – Niagara Falls

Niagara Falls is a honeypot site. Over six million visitors each year travel to see and hear 27,276 m³ of water per second plummeting over the falls. The area has been filled with hotels and restaurants to cater for these tourists and many people earn a living solely from tourism.

Honeypot site: a place of special interest that attracts many tourists and is often congested at peak times.

Figure 4 Niagara Falls

Figure 5 Niagara Falls sketch

Go Active

1 Make a copy of Figure 5 in the middle of a piece of paper. Label on it the Falls, a hotel, a restaurant and a road.

2 In one colour, annotate around the sketch to explain how these features will attract visitors.

3 In a different colour explain how the number of visitors may cause problems.

Example 2 – Flooding

Case Study – Boscastle, Cornwall 16 August 2004

Causes

- Ground was waterlogged from previous rainfall.
- Rocks in the area are impermeable.
- Summer storms brought 200 mm of rain in four hours.
- Steep slopes and narrow valleys meant that water reached the Valency river very quickly.
- Bridges trapped branches carried in the water. They quickly became blocked and acted like dams.

Impact

- 100 people had to be airlifted to safety.
- 80 cars were washed out to sea and properties were flooded.
- The flood caused £300 million of damage.

Go Active

Look back through your lesson notes at a case study of a flooding event in a **Less Economically Developed Country (LEDC)**, such as Bangladesh. Compare the causes and impacts of the floods in your chosen country with the example of Boscastle. Record your thoughts in a diagram such as the one below. Put the criteria you are comparing in the circles down the centre of the diagram and the differences in the outer circles. Think about reasons for the differences.

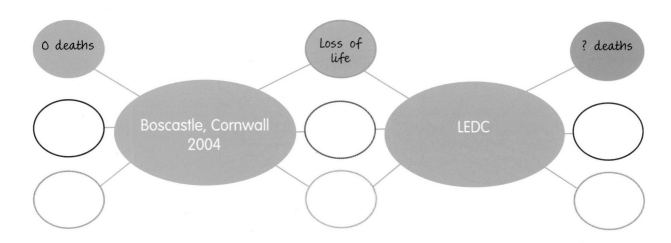

The Essentials

Causes

A flood occurs when a river overflows its banks. Floods have physical and human causes.

Go Active

1   Make a list of the physical and human causes of flooding.

2   A flood hydrograph looks at the relationship between precipitation and river flow. Look at page 10 of the Core textbook to remind yourself of what a hydrograph looks like. Give definitions for each of the terms in the table below. Imagine these definitions are to be used by someone who hasn't studied this topic. How could you simplify them?

Feature	Definition
Lag time	
Peak discharge	
Storm flow	
Rising limb	
Base flow	
Falling limb	

How should rivers be managed?

How successful are different management approaches in combating the problem of flooding?

Management

Methods used to prevent river flooding can be divided into hard engineering and soft engineering approaches.

Hard engineering uses machinery or defences constructed by people to control natural processes, e.g. building a dam in the upper river valley. This creates a reservoir which stores water at times of heavy rainfall and controls the flow of the river. It is effective at controlling the river and is often multi-purpose, providing HEP and recreational lakes. Hard engineering is also expensive to build and may result in good farmland being destroyed and beautiful countryside spoiled.

Other hard engineering approaches to flood management include building artificial levees and the straightening and deepening of river channels.

Soft engineering works *with* the environment, rather than trying to control it. Examples of this include the restriction of new building in areas at risk of flooding, afforestation and 'ecological flooding'. This last method allows the river to flood naturally in rural areas, preventing flooding in urban areas.

Methods of flood control can also be divided into **short term** emergency measures, such as evacuation and the use of sandbags, and **long term** prevention and control measures.

The Environment Agency issues **flood warnings** to give people time to flood-proof their homes, move contents upstairs or to have the area evacuated.

Go Active

Imagine you are a news reporter following up on the Boscastle floods. Write a news flash (which should take no longer than three minutes to read out) which summarises the flood management approaches implemented in the village since 2004. Remember to include the different types of flood management, whether they were short or long term approaches and an evaluation of their success.

Should we change our approach to river and floodplain management in the future?

Planners have the task of deciding how to respond to future flood hazards. There is a debate between hard and soft engineering: prediction against reaction.

Inside Information

a) Explain how different processes over a long period of time lead to the formation of oxbow lakes. You must draw labelled diagrams as part of your answer. [6]

This type of question, worth 6 marks, demands a longer answer and will be marked using a level mark scheme. It is important that you identify the command word and key words in the question, and that you plan your answer before beginning to write.

The command word in this question is **explain** and the key words are **processes** and **oxbow lakes**. A levels mark scheme will look at the overall quality of your answer and looks like this:

Level	Explanation
Level 1: 1–2 marks	A mainly descriptive answer with limited knowledge and explanation.
Level 2: 3–4 marks	Answer demonstrates understanding, with some explanation linked to the correct sequence of change from meander to an oxbow lake.
Level 3: 5–6 marks	Answer demonstrates understanding, with explanation of how named processes in the correct sequence change from meander to an oxbow lake.

An oxbow lake is formed due to erosion of the river bank around a bend. After a while, the river will break through. An oxbow lake is also known as a cut off lake.

The river can now go two different ways but normally, unless heavy rain or a flood is witnessed, will go in the direction shown by the arrows in the diagram, meaning the oxbow lake will dry up.

Erosion – wearing away of the land by material carried by rivers, glaciers, waves and the wind.

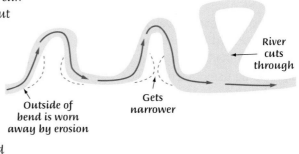

Figure 6 Student answer

Examiner's comments

In the student answer above there is understanding and a description of the correct sequence of change from a meander to an oxbow lake. However, the candidate does not identify named processes and hence this answer is marked as Level 2, awarding the candidate 4 marks.

EXAM SPOTLIGHT

Now it's your turn!

Think about the levels mark scheme above and see if you can give a Level 3 answer to these questions:

a) Draw annotated diagrams to explain how the nature of the rock and processes of erosion lead to changes in the position of a waterfall over time. [6]

b) Outline the measures that can be taken to reduce the risk of flooding. Give the advantages and disadvantages of any one of these measures. [6]

What are the causes and evidence for climate change?

What is the greenhouse effect and how have people's actions affected this process?

The Essentials

The **greenhouse effect** is the natural process by which the atmosphere traps heat from the sun. Without this process life would not be possible on Earth. It works like this:
- Solar energy reaches the Earth as shortwave radiation (sunlight).
- The Earth's surface absorbs and re-radiates the solar energy as longwave radiation (heat).
- Greenhouse gases including water vapour, carbon dioxide, nitrous oxide and methane absorb the longwave radiation, which in turn heats the lower layer of atmosphere around us.

Carbon dioxide is one of the most important greenhouse gases which regulate the temperature of the Earth.

Go Active

The carbon cycle concept map

1 Draw a line between the words in the circles which you think are connected.
2 Write on the line to explain the connection within the context of the carbon cycle.

rainwater

sunlight

leaves

earthworms

photosynthesis

respiration

carbon dioxide

decompose

carbon

atmosphere

respire

oxygen

tree

river

Climate change is a natural phenomenon.

- Past changes in the Earth's orbit around the sun have created warmer and cooler periods. Europe went through a mini ice age between 1300 and 1800.
- The output of solar radiation is not constant. Scientists have discovered that sunspots on the surface of the Sun seem to cause an increase in the temperature of the Earth.
- Volcanic activity adds to dust in the atmosphere that absorbs more sunlight.

However, there is growing evidence to suggest that people's actions are causing an increase in global temperatures. This is known as **global warming**.

Climate: this is where weather readings are taken over a long period of time (at least 30 years) and an average is taken.

15

Figure 1 Sources of greenhouse gases

Go Active

Greenhouse gases occur naturally but human activities have increased the natural amounts of these gases in the atmosphere. Use the diagram above and your own knowledge to identify, in the spider diagram, how the actions of people are increasing the greenhouse gases in the atmosphere.

Global warming: describes and explains the pattern of increasing global temperatures.

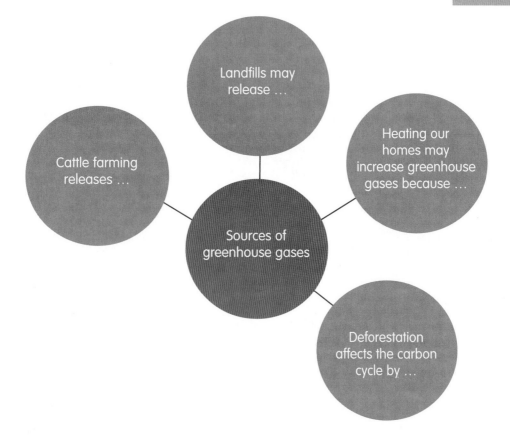

How conclusive is the evidence of climate change?

The Essentials

The evidence for global warming is considerable and growing.

- The average global temperature has increased by 0.6 °C over the past 100 years.
- Glaciers and ice sheets are melting and retreating, as with the Rhone glacier in Switzerland. This has retreated by 2.5 km over the last 150 years.
- Ice cores from the Antarctic ice sheet show that amounts of CO_2 and methane in the atmosphere have increased.
- Extreme weather events are becoming more frequent, e.g. an increase in the number and intensity of tropical storms. In recent years, the UK has experienced some of the wettest, windiest and driest periods of weather since records began.
- Bird, fish and insect species found in the climates of Africa are spreading north into Europe.

Although the evidence is persuasive, not all scientists agree that global warming is the result of human action. Some will point to evidence in the Antarctic ice cores which shows that, due to natural reasons, CO_2 levels have gone up and down throughout history although they have never been as high as they are now. The Earth's climate has always gone through natural cycles of **glacials** and **interglacials**.

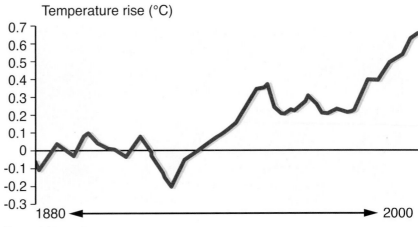

Figure 2 It's getting warmer

What are the alternative futures?

What could be the effects of climate change in MEDCs and LEDCs?

The Essentials

The consequences of global climate change could be significant and may change the way people live on our planet. Examples of the changes we can expect include:

- Sea levels could rise by as much as 1 metre by the end of the century.
- Ice caps and glaciers could continue to melt.
- Areas of the world such as Alaska and Greenland may support agriculture and life.
- Extreme weather events such as hurricanes are likely to occur more often.
- Diseases and insect pests are likely to spread into areas they have not been seen before, e.g. malarial mosquitoes may spread across Africa, Europe and the USA.
- Drought and desertification will spread. By 2020, 250 million people in Africa could face water and food shortages.
- Plant and animal species may become extinct or migrate to new areas of the world.

 Complete the spider diagram to show some possible positive consequences of global warming for the UK.

Oranges and peaches can be grown in southern Britain

Positive effects of global warming for the UK

Case Study – Rising sea levels

The maps in Figures 4 and 5 show the possible consequences of rises in sea level on the UK, a MEDC, and Bangladesh, a LEDC.

Key

Land at 5m elevation liable to flood

N

Figure 3 Rising sea levels in the UK

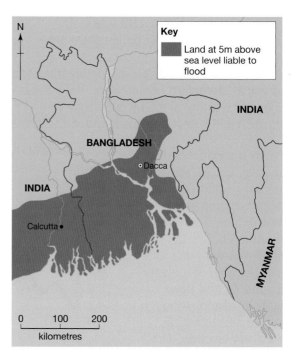

Key

Land at 5m above sea level liable to flood

N

INDIA

BANGLADESH

Dacca

INDIA

Calcutta

MYANMAR

0   100   200
kilometres

Figure 4 Rising sea levels in Bangladesh

Go Active

Imagine you are a reporter for the satellite TV channel National Geographic. Write the outline of a script which you would use to describe the possible consequences of global warming for LEDCs and MEDCs. Remember the following:

- Use the information on the maps (Figures 3 and 4) to make your descriptions as specific as possible.
- Make sure you infer at least two possible consequences per country.
- Remember that the consequences may not be directly shown on these maps. There could, for example, be a lack of food in certain areas of Bangladesh because the agricultural land has been flooded.
- Your script should last around one minute per map.

How can technology be used and people's lifestyles changed to reduce the impact of climate change?

If we accept the notion that people are making a major contribution to climate change, many argue that we need to take action to slow this change and its impacts. The causes of climate change can be tackled at a local, national and global scale.

Reducing the impact of global warming

As well as the **causes** of global warming, it is important to **reduce** its impact. Here are some ways in which this can be achieved:

• Defend and protect low-lying coastal areas from flooding.
• Improve water supplies and use them more efficiently, particularly in areas that are becoming drier.
• Encourage research into crops that will withstand drought and diseases.
• Develop more efficient ways of predicting and preparing for extreme weather events such as tropical storms.

Case Study – Kyoto Protocol

The Kyoto Protocol is an international agreement for 37 industrialised countries and the European community, which sets binding targets to reduce greenhouse gas emissions. The Kyoto Protocol was formulated in Kyoto, Japan on 11 December 1997 and enforced on 16 February 2005.

Go Active

The mind map in Figure 5 summarises some of the impacts of global warming.

1 Draw two further mind maps to summarise the causes of global warming and the measures that could be taken to reduce its impact.

2 You could laminate your mind map – use it as a table mat and revise during every meal without even thinking about it!

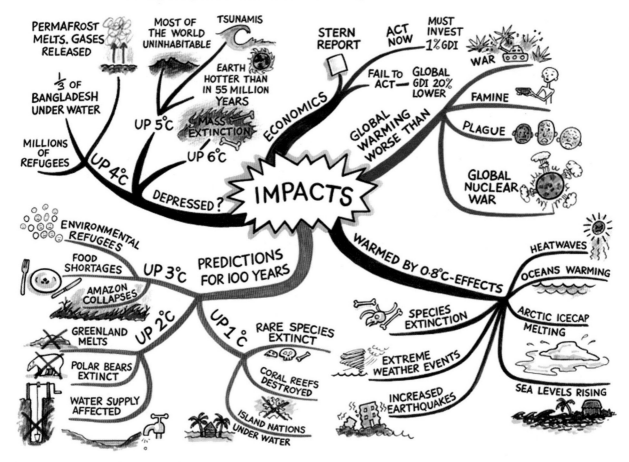

Figure 5 The impacts of global warming

Inside Information

Knowing your case studies and being able to write longer answers is important. However, shorter questions are also key to exam success because there are more of them. Generally, questions worth less than six marks will be marked using a **'point' mark scheme**. You will score one mark for every relevant point, so for shorter questions, make sure you develop your response. This way, you will gain one mark for stating the answer and one for explaining your reasoning. When developing your answers, remember these three Es:

- Give Examples
- Expand points made
- Explain the point

Questions

a) What is global warming? [2]
b) Explain one way in which human actions may cause global warming. [3]
c) Describe the possible positive effects of global warming on the lives of people in Britain. [4]

Student answers

a)

> Global warming is the way in which temperatures are rising✓ around the world. Scientists think this is the result of people burning fossil fuels✓ which adds carbon dioxide✓ to the atmosphere.

b)

> Burning fossil fuels✓ has increased levels of global warming. An example of this is people driving their cars✓ instead of walking.

c)

> Due to global warming temperatures have risen, meaning people in Britain will have hotter summers✓ to enjoy outdoors. Different plants will be able to grow. The whole of Britain will benefit, as the climate will allow all sorts of crops, plants and trees to grow. Farmers will have a longer growing season✓ and be able to grow more crops. Income for farmers will increase✓ as they will have more produce to sell.

Examiner's comments

a) An excellent answer. A relevant point is followed by two points of explanation.
b) This answer is too brief. A relevant point is followed by an example, thus scoring two marks.
c) This answer is weak. The points made are very general, e.g. 'different plants will be able to grow' is very vague and not clear enough for a mark. 'Hotter summers' is a weak point and only just worth a mark. The point made about a longer growing season is clear and relevant and therefore worth a mark. This point is then expanded, 'income for farmers will increase', earning another mark.

EXAM SPOTLIGHT

1 Now it's your turn. Rewrite the answers to questions b) and c) above to score maximum marks.
2 Study Figures 7a and 7b.
   i) Describe and explain the changing amount of carbon dioxide in the atmosphere. [4]
   ii) Compare the trend for carbon dioxide in the atmosphere with the average global temperatures between 1860 and 2000. [3]

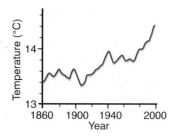

Figure 7a Average global temperatures

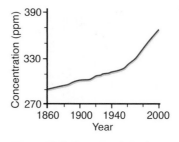

Figure 7b Carbon dioxide in the atmosphere

A The Physical World
Theme 3: Living in an Active Zone

Why are plate margins hazardous?

What are plate margins and how does plate movement generate a variety of landforms?

The Essentials
The structure of the Earth
Plates

- The Earth is made up of four distinct layers, as shown in Figure 1.
- Continental crust carries land and oceanic crust carries water.
- The crust is broken into huge slabs called plates.
- There are seven large plates and twelve smaller plates.
- Radioactive decay in the core causes heat to rise and fall inside the mantle, which in turn creates convection currents. These currents move the plates. Where convection currents diverge near the Earth's crust, plates move apart. Where they converge, plates move towards each other.
- The movement of the plates and the activity inside the Earth is called **plate tectonics**.
- The point at which two plates meet is called a **plate boundary** or **margin**. Earthquakes and volcanoes are most likely to occur near plate margins.

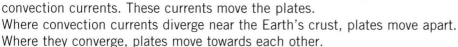

Figure 1 The structure of the Earth

Go Active

1 Describe the features of the different layers in Figure 1.
2 Label the plates A–I on Figure 2 using the list below:

> Pacific plate   South American plate   Indo-Australian plate
> North American plate   Antarctic plate   Nazca plate
> Eurasian plate   African plate   Caribbean plate

3 On a copy of your labelled map, cut out the plates and move them according to their natural direction of movement, e.g. towards, sideways or away from each other. Then, stick this new map back together.
4 When you stick the map back together, use a different coloured pen to show whether the plate boundaries are destructive or constructive margins (see below for more information on the different types of margins).

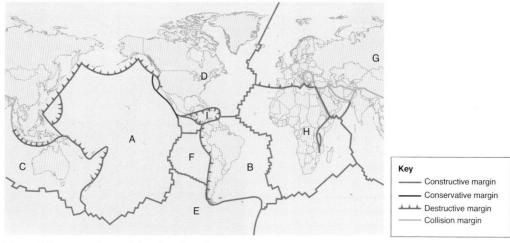

Figure 2 Plate margins and the direction of movement

21

Plate margins

Destructive plate margin

- This is where plates push together.
- The denser oceanic crust, made of basalt, is forced down into the mantle (**subduction**). This forms an **ocean trench**.
- The lighter continental crust, made of granite, is compressed to form **fold mountains** such as the Andes. As the oceanic plate sinks, it melts and the molten magma finds its way to the surface in explosive **volcanoes** such as Cotopaxi.
- **Earthquakes** occur when the plates move.

Constructive plate margin

- This is where plates move away from each other.
- Where two oceanic plates move apart, ocean trenches are formed and undersea volcanoes form mid-ocean ridges, such as the Mid-Atlantic Ridge.
- Where continental plates pull apart, **rift valleys** are formed and lava erupts to create **shield volcanoes**, such as Skjaldbreidur in Iceland. Minor earthquakes can occur here.

Conservative plate margin

- This is where plates are pushed in different directions, e.g. the San Andreas Fault.
- Plates remain locked together until the rock breaks along a fault line. Major earthquakes occur as stored energy is released.

Go Active

1 Label Figures 3 and 4 using the features listed below.

    volcanic island   explosive volcano   ocean trench   mantle   magma
        fold mountains   oceanic crust   ocean ridge   continental crust

2 Carry out your own research of the Great Rift Valley in Africa. Draw an annotated diagram to explain how this natural feature was formed.

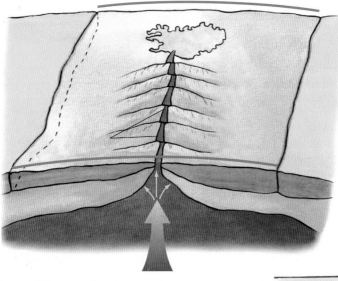

Figure 3 Constructive plate margin

Figure 4 Destructive plate margin

What are the primary and secondary hazards associated with volcanoes and earthquake zones?

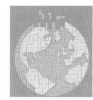

The Essentials
- Volcanoes occur where weaknesses in the Earth's crust allow magma, gas and water to erupt onto the land and seabed.
- Most volcanoes occur at plate margins.
- Hot spots are places where the crust is thin and magma may be able to reach the surface of the earth, e.g. Hawaii.
- Volcanoes can be classified into the following categories:
  Active – volcanoes that have recently erupted.
  Dormant – volcanoes that have not erupted for a long time.
  Extinct – volcanoes around which there has been no record of tectonic activity.
- Another way of describing volcanoes is according to their shape and composition.
  Shield volcanoes are found at constructive plate margins and above hotspots. Here, the lava is hot and runny, flowing long distances before solidifying. Shield volcano eruptions are gentle oozings of lava, which form large cone shaped mountains, like Mauna Loa in Hawaii and Surtsey in Iceland.
  Composite volcanoes are found at destructive plate margins. Here, lava is acidic, thick, sticky and cools quickly. Composite volcanoes experience explosive eruptions of ash, lava and lava bombs, which are cone shaped with steep sides, like Montserrat in the Caribbean and Mount St Helens in the USA.

Go Active
Label Figure 5 with the following terms:

1 Ash
2 Lava
3 Vent
4 Magma chamber
5 Crater
6 Secondary cone
7 Lava bomb

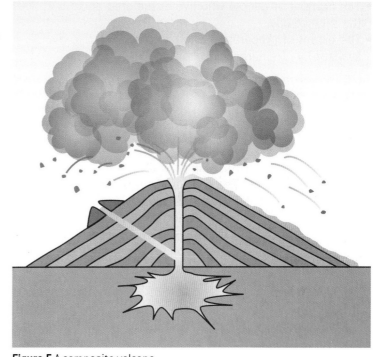

Figure 5 A composite volcano

Volcanic hazards

Hazard	Description
Lava flow	Molten rock flowing down the sides of a volcano. Hot basaltic lava from shield volcanoes flows quickly.
Lahars	Mudflows, a mixture of ash and water from melted snow and ice, travel at great speed down the mountain, making evacuation difficult.
Dust and ash clouds	Ash thrown high into the atmosphere shuts out the sun and, when it settles, can completely bury buildings and crops.
Lava bombs	Large pieces of rock and ash are thrown into the air.
Pyroclastic flow	Burning clouds of gas and ash, with temperatures up to 1000°C, rush down the mountain, scorching everything in their paths.

Earthquakes

- An earthquake is the result of vibrations in the Earth's crust. These are caused by shock waves travelling outwards from a sudden movement deep within the crust.
- The source of the shock wave is known as the **focus**, and the point on the Earth's surface immediately above is the **epicentre**.
- Most large earthquakes are associated with movements along plate margins but many smaller movements occur at weaknesses in the crust, such as fault lines.
- The size of an earthquake is measured with a **seismograph** along the **Richter scale**.

Earthquake hazards

The impact of an earthquake will depend on a number of factors:

- The strength of the earthquake and distance from the epicentre.
- The nature of surface rock. Some rocks 'shake' more than others.
- The number of people who live in the area and the time of day.
- The extent of preparation in an area and availability of emergency services.

Tsunamis are powerful ocean waves caused by earthquakes. They travel long distances across oceans very quickly. As they approach shallower water near the coast, the wave builds up to great heights. On 26 December 2004, an earthquake off the west coast of Sumatra, measuring 9.1 on the Richter scale, triggered a tsunami 30 metres high. It killed over 200,000 people and left two million people homeless in low-lying coastal areas.

Primary impacts: these are the immediate consequences of the event, e.g. earthquakes cause buildings to collapse and damage to roads/bridges/railway lines.

Secondary impacts: these are consequences which result from the primary impacts, e.g. buildings collapsing lead to homelessness, ash clouds from volcanic eruptions ground aircraft.

Case Study – The 2008 Sichuan Earthquake

Fact file

- Sichuan is in central China.
- The earthquake registered 8.0 on the Richter scale.
- The area is densely populated and the earthquake struck at 4 a.m., when most people were asleep.
- It was caused by movement of the Indian plate against the Eurasian plate, the same movement responsible for the creation of the Himalayas.

Primary effects

- Buildings collapsed and many thousands were buried under rubble.
- 70,000 people were killed.
- 5 million people were left homeless.
- 374,000 people were injured.
- Many farm animals were lost and crops destroyed.

Secondary effects

- Relief efforts were hampered because roads were destroyed.
- 700 schools were destroyed.
- Water and electricity supplies were cut off.
- Fires broke out in the collapsed buildings.

Responses

- 50,000 army troops were sent to help.
- International aid and equipment were brought in, including earthquake experts from Japan.
- 1.47 million people were evacuated from the worst affected areas.
- Temporary camps were set up with 34,000 tents erected.
- The government passed strict building laws and regulations to help prevent future damage.

Go Active

Sort the following secondary impacts of volcanic and earthquake hazards into economic, social or environmental categories. You can record your answer any way you like, e.g. by using a table or colour coding.

People left homeless	Gas mains fractured	Shops run out of supplies	Bridges destroyed
Crops destroyed	Roads blocked	Fish die in local rivers	Soil fertility improves
Disease breaks out	Increase in tourism	Schools close	Ash blocks sunlight
Tsunami hits coastline	Factories closed	Trees flattened	

Why do people continue to live in hazard zones?

The Essentials

500 million people live in active zones despite the hazards. Here are some reasons why:

- The dramatic scenery created by volcanic eruptions attracts tourists. This brings income to an area, e.g. Iceland and Mount Etna.
- The lava and ash deposited during an eruption breaks down to provide valuable nutrients for the soil. This makes it very fertile which is good for agriculture.
- The high level of heat and activity inside the Earth close to a volcano, can provide opportunities for generating energy. This is called **geothermal energy**.
- Volcanic rock makes good building stone.
- Many people live in such places because they were born there and do not want to move.
- Some people, particularly in LEDCs, cannot afford to move.
- Seismic events are often rare and may not affect an area in a person's lifetime.
- **Prediction** and earthquake resistant buildings are improving all the time.

Case Study – Montserrat

Fact file
- Montserrat is an island in the Caribbean with a population of 11,000.
- Main eruptions took place between 1995 and 1997 and have continued.
- Montserrat lies on the destructive plate margin between the North American and Caribbean plates.

Primary effects
- 19 people died.
- Pyroclastic flows burned buildings and trees.
- Ash buried over two thirds of the island.
- 60 per cent of housing was destroyed in the capital city, Plymouth.
- The airport was buried in ash and roads destroyed.

Secondary effects
- Hospitals and schools closed.
- Farming became impossible, as fields were buried under ash.
- Coral reefs were destroyed by ash being washed into the sea.

Responses
- The volcano had not erupted for 400 years and the government had no plans in place for such an event.
- People were evacuated to the north of the island.
- 8000 people fled Montserrat, as refugees, to the UK and neighbouring islands.
- The UK had given £55 million in aid by 2010.
- Little farmland has been reclaimed in the south.
- Plymouth remains uninhabited.
- Tourism is slowly returning – the volcano is now a tourist attraction.

Go Active

After each case study, make a case study card. The card should be A5 or smaller and should contain the essential information. This information should be in written and picture form, as pictorial representation has been proven to aid memory, and must include at least some facts and figures. You could use the following as a guide:

Name of case study
Unit/topic to which it relates
LEDC or MEDC?
Location and location map
Reasons for case study
Effects of case study, including immediate, short and long term, local, national and global and environmental, social and economic
Consequences of or solutions to case study

All information must fit onto one card. You should keep these cards safe, as they can be used to revise immediately before the exam.

How can the risks associated with volcanic and earthquake zones be reduced?

There is nothing that can be done to stop volcanic eruptions or earthquakes; prevention is not an option. This leaves two possible ways of managing hazards such as earthquakes and volcanoes: prediction and preparation.

The Essentials

Predicting eruptions

Volcanologists (people who study volcanoes) use a variety of monitoring techniques to predict eruptions:

- **Remote sensing** – satellites monitor volcano temperature and gas emissions.
- **Seismometers** – these measure the increase in earthquake activity that occurs before an eruption.
- **Tiltmeters** – these monitor changes in the shape of a volcano that occur as it fills with magma.
- **Gas emissions** – indicate increased risk.
- **Ultrasound** – this is used to detect the movements of magma.

Figure 6 Molten lava pouring from an erupting volcano

Preparing for eruptions

A detailed plan is needed for dealing with a possible eruption. The following elements must be considered:

- An **exclusion zone** around the volcano must be created.
- Authorities must be ready and able to evacuate residents.
- Emergency supplies of basic provisions such as food must be gathered.
- A good communication system needs to be in place.
- Everyone who could be affected by the eruption needs to be informed.

Predicting earthquakes

Earthquakes are not as easy to predict as volcanic eruptions but there are still some ways of monitoring activity:

- Laser beams can be used to detect **plate movement**.
- A seismometer is used to pick up vibrations in the Earth's crust. An increase in vibrations may indicate an earthquake.

- **Radon gas** escapes from cracks in the Earth's crust before an earthquake. Levels of radon gas can be monitored – an increase may suggest an earthquake.

Preparing for an earthquake

The prediction techniques used to monitor earthquakes are not 100 per cent reliable. Therefore, preparation is vital:

- People living in earthquake zones need to know what they should do in the event of a quake. Training people may involve holding earthquake drills and educating people via TV or radio.
- People may put together emergency kits and store them in their homes. An emergency kit may include first aid items, blankets and tinned food.
- Earthquake proof buildings have been constructed in many major cities, e.g. The Transamerica Pyramid in San Francisco. They are designed to absorb the energy of an earthquake and to withstand the movement of the Earth.
- Roads and bridges can also be designed to withstand the power of earthquakes.

Go Active

1 Choose **four** things people could do to prepare for an earthquake and **two** ways in which earthquakes can be predicted.

2 Now draw and label a picture for each of the six things on six different 'Post-it' notes. Stick them to your bedside table and have a quick look at them every night and every morning. This will help you to remember them. It would be even more useful if you had one colour to indicate preparing for earthquakes and one for predicting them.

Inside Information

Sample question
Annotate the diagram to explain why volcanic eruptions occur at plate margins. [4]

Student answer

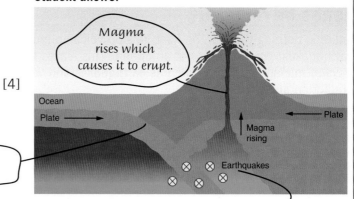

Magma rises which causes it to erupt.

Ocean

Plate →

Oceanic crust is destroyed.

Plate ←

Magma rising

⊗ ⊗ ⊗ Earthquakes
⊗ ⊗ ⊗

Friction increases heat, crust is turned into magma.

Examiner's comments
Annotate is a command word which demands that candidates add explanatory notes. It requires more than labelling, which only asks for one word or short phrase.

The student answer shows good understanding of both the instruction to annotate and the processes taking place at this plate margin. However, there is not enough detail here for four marks, as the candidate describes rather than explains the events. Friction is identified as the force which creates the heat that melts the oceanic crust, but I am left wondering what force is causing the oceanic crust to be pushed under the continental crust and why is the magma rising.

I would give this one mark for 'Oceanic crust is destroyed' and one mark for 'Magma rises which causes it to erupt'. This gives a total of two marks.

EXAM SPOTLIGHT

Study the diagram below, which shows the distribution of volcanoes.
a) Give two facts about the distribution of volcanoes. [2]
b) Mount St Helens is an active volcano located on a destructive plate boundary. Describe how volcanoes such as Mount St Helens can be monitored with the aim of predicting the next eruption. [4]
c) Earthquakes are another example of tectonic activity. With reference to an example you have studied, describe the primary and secondary effects of an earthquake. [6]

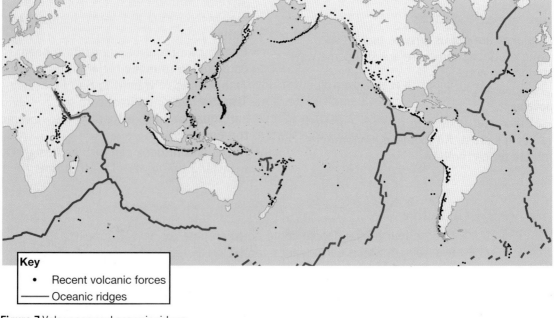

Key
• Recent volcanic forces
— Oceanic ridges

Figure 7 Volcanoes and oceanic ridges

Where do people live?

Where do people live in the world and why do they live there?

The Essentials

Distribution and density

People are not evenly spread over the Earth's surface.

- **Population density** is defined as the number of people living in a square kilometre. Places can be described as being densely or sparsely populated.

- **Population distribution** describes how people are spread out. Some terms used to describe distribution include even, clustered and random.

Reasons for differences in population distribution and density include:

- a range of **physical factors** such as relief, climate, soils, vegetation and natural resources.
- a range of **human factors** such as urban growth, industrial growth, agricultural development, accessibility and government policies.

Case Study – Where people live in Brazil

An excellent example of how the above factors operate can be seen in Brazil, South America.

- Brazil has a population of 186 million.
- The **north** is sparsely populated. It contains the Amazon river and the dense vegetation of the rainforest.
- The **north east** of the country contains a third of the population. Numbers are decreasing due to drought which is devastating fertile farming areas.
- The **south east** has the highest population density. It contains the cities Rio de Janeiro and Sao Paulo and is the industrial centre of Brazil.
- In the **south** of Brazil, fertile soils and a suitable climate support a prosperous agricultural region, which can sustain the high population density.
- The **centre west** is inaccessible and has a low population density. However, following the political decision to move the capital to Brasilia, the population of the area is increasing.

Go Active

Which of the following statements would you choose to describe the population distribution of North America? Why is this a good description? How could you improve it?

a) People are evenly spread over the continent.
b) People are unevenly spread over North America, although more people live in the east.
c) Parts of the south are densely populated but few people live in the north.
d) The north is sparsely populated. There are more people living in the south and east, although parts of the west coast are also densely populated.

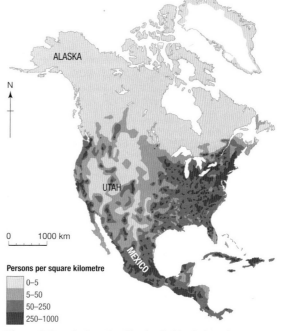

Persons per square kilometre

- 0–5
- 5–50
- 50–250
- 250–1000

Figure 1 Population distribution in North America

What are the push/pull factors that produce rural-urban migration in LEDCs and urban-rural migration in many MEDCs?

The Essentials

LEDCs (Less Economically Developed Countries), such as Uganda and Bangladesh, are poorer countries with low standards of living.

MEDCs (More Economically Developed Countries), such as France and Japan, are wealthier countries with high standards of living.

Urbanisation

This is the process by which an increasing proportion of a population live in urban areas. Migration of people from the countryside to cities, together with higher birth rates and international migration in urban areas, lead to urbanisation.

Counterurbanisation

This is a term, first used in the USA during 1970s, to describe the movement of people from urban to rural areas.

Rural-urban migration

Rural to urban migration is common in many LEDCs. This happens because:	**Urban to rural migration** is a feature of many MEDCs. This happens because:
• High population increases cause land and food shortages in rural areas. • People believe that urban areas are places of 'bright lights', job opportunities and access to services such as schools and hospitals. • Crop failures, lack of money, absentee landlords and a lack of education mean a life of poverty in the countryside. • Farm mechanisation causes unemployment and there are few alternative jobs outside the city.	• Urban areas are increasingly seen as places of noise, crime and pollution. • Rural areas are seen as places where there is space, quiet and a sense of community. • Rural areas are thought to have good quality schools and be a safe place for children. • Increased wealth and car ownership allow easy access to urban areas. • Improvements in telecommunications make it possible for people to work from home. • New business parks are built at the urban-rural fringe so people no longer need to commute to the city centre. • Many people retire to the countryside.

Go Active

1 Fill in the boxes below to explain reasons for and the effects of people migrating from an urban area to a rural area in an MEDC. Add a new row of boxes for each new reason.

Reason	Immediate/short term consequence	Long term consequence	Any other consequences?

2 Now do the same for rural-urban migration in LEDCs.

What will happen to the world population?

What are the factors that influence birth rates and death rates?

The Essentials
Global population growth

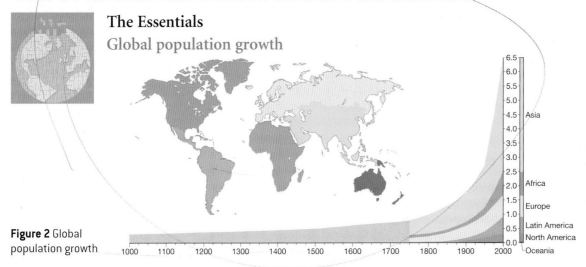

Figure 2 Global population growth

- The global population increased slowly until around 1800. Since then, world population growth has increased rapidly.
- In 2000 the world's population was estimated to be 6 billion, double that of the 1960s. The majority of this recent growth has taken place in LEDCs.
- Today, social and economic changes in LEDCs have lead to slower population growth, while improvements in agriculture have reduced food shortages.
- Projections of further growth vary. Most experts think that world population will reach 8 billion this century.
- Global population growth depends on a balance between births and deaths. **Natural increase** is the difference between the birth rate and the death rate.

Go Active

Study Figure 2. Which of the following four statements would you choose to describe world population growth? Why did you choose this statement over the others?

a) Every continent has experienced a population growth in the last 1000 years.

b) Every continent has experienced a population growth in the last 1000 years but Asia has experienced the biggest growth.

c) Every continent has experienced a population growth in the last 1000 years but Asia has experienced the biggest growth. The population in Asia had increased to around 4 billion by the year 2000.

d) Every continent has experienced a population growth in the last 1000 years but Asia has experienced the biggest growth. Europe and Africa have experienced similar rates of population growth, with Africa's being a little larger and a little faster.

Birth rate: the number of live births per 1000 people per year.

Birth rates

Birth rates are higher in most LEDCs because:
- children provide labour on farms and security for old age
- large families are seen as a sign of male virility
- some religions do not approve of contraception
- girls marry early, extending their child bearing years
- women may lack education and be expected to stay at home to raise a family
- availability and knowledge of contraception may be limited in some rural areas
- a high infant mortality rate encourages large families.

Birth rates tend to be lower in MEDCs because:
- people marry later
- women are educated, often have careers and delay the start of a family
- the high cost of living means it is expensive to raise children
- couples prefer to spend money on material things such as holidays and cars
- birth control and the contraceptive pill in particular, are easily available.

Death rates

Death rates are low in MEDCs and falling in LEDCs because:

- better health care is more widely available to people
- people have less physically demanding jobs
- methods of preventing diseases such as malaria, cancers and cholera are being developed
- people are better educated about health and hygiene
- water supplies are more reliable and cleaner
- there is more sanitary disposal of waste
- agricultural improvements and higher incomes provide better food supplies and living conditions.

Death rates are increasing in some MEDCs and LEDCs because:

- HIV is having an increasingly significant impact on death rates in LEDCs.
- In MEDCs there are an increasing number of elderly people.

> **Death rate**: the number of deaths per 1000 people per year.

How do differences in birth and death rates affect population numbers and structures in South Asia, sub-Saharan Africa and Western Europe?

The Essentials

Population structure

Population structure is displayed in a population pyramid.
- Population is divided into five-year age groups.
- Horizontal bars show the percentage of the population in each age group.
- Males are shown on the left and females on the right.

Population structure is often divided into three age groups:
- Young dependents, 0–14
- Working population, 15–64
- Old dependents, 65+

Population structure will change with development. The higher birth rates in LEDCs typically produce population pyramids with a wide base. However, the lower birth rates in MEDCs produce population pyramids with a narrow base and wider top.

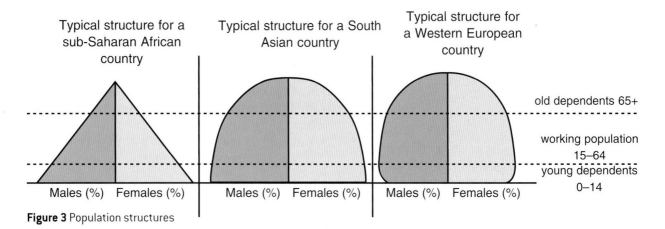

Figure 3 Population structures

Go Active

1. Look at Figure 3. Compare the number of dependents in these three parts of the world. Focus on:

    a) the number of young dependents

    b) the number of old dependents with increasing development.

2. Explain how the changing population structure will affect the demand for services in a country.

Case Study – Nigeria (sub-Saharan Africa)

Sub-Saharan Africa is a geographical term used to describe those African countries which are fully or partially located south of the Sahara. This area contrasts with north Africa, which is considered part of the Arab World.

Nigeria is a sub-Saharan country located in west Africa and is the continent's most populous country, with a population of 148 million. Nigeria has a rapidly growing economy but is a typical LEDC.

Look at Figure 4. The wide base shows that Nigeria has a high birth rate, while the narrow top shows there are fewer older people in Nigeria's population. The shape

is typical of a sub-Saharan African country with a wide base and a narrow top.

However, the consequences of a high birth rate mean there are a lot of young children who need to be fed, housed, educated and have medical care provided for them. This puts a lot of pressure on Nigeria's services.

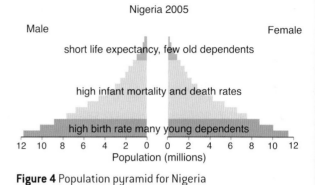

Figure 4 Population pyramid for Nigeria

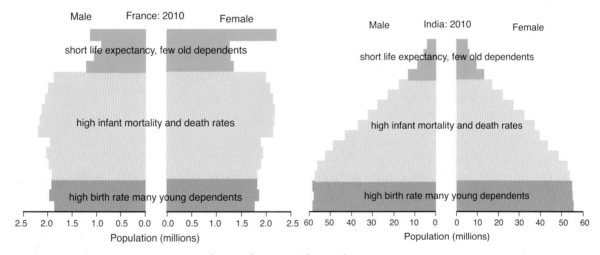

Figure 5 A population pyramid for France (a MEDC) and India (a LEDC)

Go Active

Look at Figure 5. Use the double bubble below to describe how the population structures in France and India differ in 2010. You have been given three different areas of the population pyramid to focus on. Remember to show the examiner that you have looked at Figure 6 by quoting figures in your descriptions.

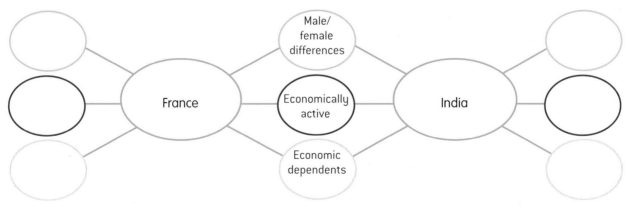

How might these differences change in the future?

The Essentials

Rates of population growth vary across the world. Although the Earth's total population is rising rapidly, not all countries are experiencing this growth:
• **MEDCs** generally have low population growth rates.
• **LEDCs** generally have high population growth rates.

MEDCs	Birth rates	Death rates	Fertility
UK	12	10	1.6
Germany	8	10	1.3
Italy	10	9	1.4

LEDCs	Birth rates	Death rates	Fertility
Ghana	33	10	4.4
Kenya	40	12	4.9
India	24	8	2.9

Fertility: the average number of children in a family.

Population change in MEDCs

Birth rates are falling in most countries around the world and improvements in health care, nutrition and living conditions mean that people are living longer. In these conditions, the challenge that many governments face, particularly in MEDCs, is how to support an increasingly old population.

Case Study – The UK's aging population
What are the issues?
• There are decreasing numbers of economically active people in the population.
• There are more elderly dependents.

What can be done about it?
• People are being encouraged to save for their retirement.
• People are working for longer: the age at which people retire is increasing.
• Facilities such as nursing homes and care workers are needed, perhaps in preference to schools, as the population gets older.
• Educated and skilled migrants could be encouraged into the country to fill labour shortages.
• The country could adopt a pro-natalist policy. This would encourage people to have more children by offering them benefits, such as better access to childcare and better conditions for maternity leave.

Population change in LEDCs

In many countries around the world, how to manage rapid population growth is an important issue but in some, particularly those in sub-Saharan Africa, the AIDS epidemic is claiming the lives of young and fit members of the working population. This distorts the population structure and is of pressing concern.

Case Study – China

During the late 1970s and in an attempt to slow down the rate of population growth, the Chinese government introduced their famous one child policy.

- The policy, passed in 1979, meant that each couple in China was allowed just one child.
- Benefits, including access to education and health care, were offered to families that followed this rule and withdrawn from those households with more children.
- Fines were placed on families that had more than one child.
- The one child policy was resisted in rural areas, where it was traditional to have large families.
- Gender-selective terminations (where baby girls are aborted) have become common.

Impact of the policy

- The birth rate in China has fallen considerably in the last 30 years. The rate of population growth is now just 0.7 per cent, compared with 1.9 per cent in the 1970s.
- Large numbers of female babies have ended up homeless or in orphanages. In 2000, it was reported that 90 per cent of foetuses aborted in China were female.
- China's gender balance has become distorted. Today it is thought that men outnumber women by more than 60 million.

Long term implications

- The falling birth rate is leading to an unbalanced population structure, with the number of elderly people rising.
- There are fewer people of working age to support the growing number of elderly dependents.

Go Active

Make a case study revision card for each of the above cases. Remember the card should be an A5 piece of card and should contain the essentials of the study. This information should be in written and picture form and must include at least some facts and figures. You could use the following headings:

Name of case study
Unit/topic to which it relates
LEDC or MEDC?
Location and location map
Reasons for case study

Go Active

Complete a PMI (Plus, Minus, Interesting) table for each of the suggestions below. Try to come up with at least three 'pluses', three 'minuses' and one 'interesting' point for each suggestion. (An 'interesting' point might be a question raised or your opinion.)

P	M	I

1 Increase population growth in MEDCs by paying people if they have three or more children.
2 Increase population growth in MEDCs by providing free IVF for lesbian couples.
3 Balance the population structure in Nigeria by reducing the birth rate and the death rate.

Inside Information

Explain why the world's population is not evenly spread around the globe. [4]

Examiner's comments

- It is vital that you understand and carry out the instructions indicated by the command words. The command word in this question is **explain**. The key words are **population** and **spread**. The question is therefore asking you to explain a distribution.

The answers to this question are marked using a points mark scheme. Marks were awarded for relevant points relating to climate, water supply, fertile soil, vegetation, relief, oceans, hazards, natural resources and industrial development.

You could also achieve marks by developing any one of these points.

Sample answers

Student A

Much of the world's population live in the richer northern hemisphere. This is because infrastructure is far greater here. The climate conditions✓ are more favourable than those of Africa or Australasia. This results in a higher level of tourism. Therefore the pull factors are great, as are the push factors in the southern hemisphere.

Examiner's comments

This is a poor answer. It is confused and far too general. One mark was given for the very weak point made about climatic conditions.

Student B

Some places are more populated than others because of things like land. If it is flat✓ it is easy to build on,✓ [development] like in Bangladesh, and this will be more populated than a bumpy terrain like that of Nepal. Also, fertile soil✓ attracts more people, as in Holland,✓ [development]. Also, places with lots of raw materials,✓ e.g. coal✓ [development] are likely to be more densely populated. Also, better climates, e.g. not too hot or too cold✓ will be more densely populated.

Examiner's comments

This is a good answer! Four clear points have been made and three of these have been developed. By making these points, and by the development, this student has scored the maximum mark of 4. It is pleasing to see real places given as examples.

Study Figure 6.

a) Cities in LEDCs like Botswana are growing because of rural push factors and urban pull factors. List three rural push factors. [3]

b) Why is there a low proportion of old people and children in this city. [3]

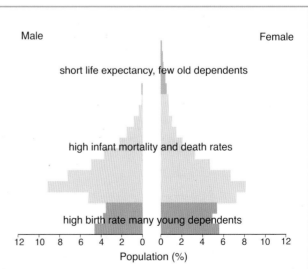

Figure 6 A population pyramid for a city in Botswana

What is globalisation?

How have changes in business and technology increased interdependence between MEDCs and LEDCs?

Globalisation: this is when human activities take place on a worldwide scale, meaning we increasingly live in a 'global village'.

Interdependence: this is when countries are linked together economically, socially, culturally and politically so that they are dependent on each other.

Go Active

Write definitions for globalisation and interdependence that could simply and easily explain the terms to a Year 7 student.

The Essentials
Globalisation

Globalisation, a term first used in the 1950s, is the result of the following:

- Improvements in technology and telecommunications – computers, internet access, email, mobile phones and video conferencing.
- Improvements in transport – people now holiday all over the world and businesses ship products and raw materials globally.
- The growth of multinational companies (MNCs,) such as HSBC and Nike. These companies are the driving force of globalisation.
- Greater political cooperation, e.g. the World Trade Organization (WTO), an inter-government organisation that promotes free trade.
- The development of trading blocs.

Multinationals (also called Transnationals)

McDonald's, the US fast food chain, is a large MNC. It has nearly 30,000 restaurants in 119 countries and has helped create a single global economy. The majority of multinationals have their headquarters in MEDCs like the USA and the UK but often locate branches in poorer countries, e.g. B&Q now has stores in China. They also invest in other MEDCs, e.g. the American company Ford in the UK.

Factors attracting multinationals to a country include:
- cheaper raw materials
- cheaper wages
- relaxed environmental laws
- good transport links
- access to the market where the goods are sold
- friendly governmental policies which offer grants and tax incentives.

Go Active

Have a look around your home and reflect on the daily impact MNCs have on your life. Complete a spider diagram or mind map which summarises these effects of globalisation. Perhaps you could divide the effects into good, bad and neutral categories.

What are the benefits of globalisation and why do some see it as a threat?

The Essentials

Benefits of globalisation

Globalisation is having a dramatic effect on the lives of people. Some positive impacts:

- Investment by MNCs provides new jobs and skills for local people.
- Multinationals bring foreign currency to local economies when they buy local resources, products and services. This is known as a **multiplier effect**.
- The mixing of people and cultures from all over the world enables a sharing of ideas and lifestyles creating a vibrant social mix. People can experience foods and products not previously available in their country. They can take holidays in distant places.
- Migration of people can fill labour and skill shortages.
- Globalisation can help make people aware of events in distant parts of the world, e.g. people in the UK were quickly alerted to the impact of the 2010 earthquake in Haiti.
- It may help make people more aware of global issues like deforestation and global warming and alert them to the need for sustainable development.

The threats of globalisation

Critics of globalisation include environmentalists, anti-poverty campaigners and trade unionists. The negative impacts they point to include the following:

- Globalisation operates largely in the interests of the richest countries. The role of the developing countries is often to provide the richer countries with cheap labour and raw materials.
- Profits are often sent back to the MEDC where most multinational companies are based.
- MNCs, with their large-scale economies, may drive local companies out of business.
- If it is cheaper to operate in other countries, MNCs might close down local factories and make people redundant.
- Multinationals may operate in a way that would not be allowed in a MEDC, e.g. polluting the environment, taking risks with safety or paying low wages to local workers.
- Globalisation is a threat to the world's cultural diversity, as it drowns out local traditions and languages, re-casting the whole world in the mould of the capitalist West.
- Migration of people across the world can cause social tensions.
- Industry may begin to thrive in LEDCs at the expense of jobs in MEDCs. The decline of traditional industries in MEDCs is known as **deindustrialisation**.

Go Active

1 When you have read through the positive impacts and the threats of globalisation, close this revision guide.
2 Then try to make two lists, one including the benefits of globalisation and one the threats. When you've made this list, open up the revision guide again and see how many you can remember.
3 Repeat the activity several times until you can remember at least three benefits and three threats. Remember, it is an advantage if you can remember some details and case studies for the examples you give.

Case Study – Health workers

From 2000 to 2005, between 10,000 and 15,000 newly trained nurses from LEDCs joined the UK's National Health Service. The advantages and disadvantages of this brain drain include the following:

- Staff earn far more than they would in their home country. Some of these earnings can be sent home.
- Staff benefit from training in the latest techniques and medical treatments. This knowledge will be taken home when staff return.
- The NHS would have serious shortages of nurses and doctors if they did not recruit from abroad.
- Waiting times are reduced and patients in the UK get faster treatment.
- Hospitals in LEDCs, particularly in Africa, are desperately short of trained staff.
- Money spent in LEDCs on university training is lost when graduates migrate abroad.

What are the impacts of globalisation on countries at different levels of development?

What have been the social and economic impacts of the enlargement of the EU?

The Essentials

The European Union

- The UK is a member of the European Union, a group of countries whose governments work together to make trade easier and improve living standards of its people. It's a bit like a club – to join you have to agree to follow the rules and in return you get certain benefits.
- The European Union has 500 million citizens. Each country pays to be a member and the money is used to change the way people live and do business in Europe.
- The EU gets rid of controls that stop people moving around freely inside the Union.
- There are some who think that the EU does too much and that it tries to create universal rules for everyone in Europe, making people obey them even if they disagree.

Date joined	Country
1957	Belgium, France, the Netherlands, Germany, Italy,
1973	Luxembourg
1981	Denmark, Ireland, UK
1986	Greece
1995	Portugal, Spain
2004	Austria, Finland, Sweden Hungary, Poland, the Czech Republic, the Slovak Republic, Slovenia, Estonia, Latvia, Lithuania, Malta, Cyprus
2007	Bulgaria, Romania

A **trading bloc** is a group of countries that work together to remove barriers and improve trade between member countries.

Impacts of the enlargement of the EU

An enlarged European Union has increased trade between member countries and made it easier to travel between countries. Tourism in cities such as Krakow and Budapest is booming and many workers in eastern Europe are travelling to the west in search of work.

Case Study – Polish migration to the UK

- People who come from countries in the EU can live and work in any other EU country. Half a million people from Poland came to the UK after it joined in 2004.
- People left Poland because of high unemployment, low wages and housing shortages.
- They came to the UK because there was more work, higher wages and a big demand for trades people such as plumbers and electricians. It was also easy to move to Britain as English is an international language and the UK allowed unlimited migration.

Impact on Poland	Impact in the UK
• Poland's population fell by 0.3 per cent. • The birth rate fell as most of those who left were young and of child-producing age. • There will be fewer children in schools in the future. • Money was sent home by migrant workers, (about £3 billion in 2006).	• The UK population increased slightly. • Immigration filled jobs and boosted the UK economy. • Significant amounts of money were sent out of the country. • New shops selling Polish products opened. • Attendance at Catholic churches increased. • There was tension in some areas between the host population and Polish migrants.

Go Active

Using the table above, colour the social impacts of Polish migration into the UK in red and the economic impacts in blue.

How have newly industrialised countries such as India and China benefited from globalisation?

The Essentials

China and India are often used as examples of the success of globalisation in improving the lives of people in LEDCs. China's market is huge and only recently, with China's entry into the WTO, is the government opening it up to world trade. China's growth has been impressive but it has created tremendous internal inequalities.

- Growth has been concentrated in the coastal regions, with the western and northwestern provinces being left behind. Inequalities among the countryside and the city, regions and classes are all growing.
- Over 120 million people have been lifted out of poverty by China's growth but there are increasing problems arising from tensions between the new classes of 'haves' and 'have nots'.
- China cannot continue on the path of high speed growth without incurring serious environmental problems.
- MNCs from the USA, Europe and Japan have moved to China, not so much to exploit a domestic market but to make it a global manufacturing base.

Advantages of MNCs in China	Disadvantages of MNCs in China
They bring jobs to Chinese people. The government can tax them and use the money to help develop China's economy. They often use materials and equipment made by Chinese companies. They bring with them new skills and technologies.	They often pay low wages. Some use children as workers in poor conditions. They may pollute the environment. They may quickly move their offices, shops or factories to other countries if it works out cheaper. Chinese culture for the worse, e.g. new fast food restaurants etc.

Case Study – The growth of manufacturing in China

- In 30 years China has gone from a mainly agricultural economy to a strong manufacturing economy, e.g. China manufactured 75 million televisions in 2004 compared with 4000 in 1978.
- China is now the third largest economy after the USA and Japan.
- China manufactures a wide range of products sold globally such as clothes, computers and toys.
- Many MNCs, such as Nike and Disney, have locations in China.

Nike is an American MNC that sells sportswear. Their research, design and marketing are based in the United States but manufacturing takes place worldwide in countries such as China and Bangladesh. Nike doesn't own factories but sub-contracts the work to factory owners. In 2007 Nike made a profit of $1.5 billion. Reasons for the growth of China's manufacturing industry include:

- Cheap labour – there is no minimum wage in China. A worker can expect to earn around £70 a month.

- Long working hours – Chinese law states that workers should work a 40 hour week and a maximum of 36 hours overtime. However, this law is not enforced and some workers endure up to 80 hours of overtime.
- More relaxed health and safety regulations – laws exist but are not heavily enforced.
- Prohibition of strikes – it is illegal for people to join any other union except the All-China Federation of Trade Unions (ACFTU). This union can go on strike but is required by law to get people back to work as soon as possible.
- Tax incentives – China has many Special Economic Zones (SEZs) that offer tax incentives to foreign businesses. Foreign manufacturers usually pay no tax for their first two years in China, 50 per cent tax for the next three years and then fifteen per cent tax. Shenzhen is one of the most successful SEZs. Factories in Shenzhen make products for companies like Wal-Mart, Dell and IBM.

How have patterns of trade hindered economic progress in the least developed LEDCs?

Imports are goods purchased from abroad and brought into a country.

Exports are goods purchased by other countries and sent to them. The balance of trade is the difference between the money earned from exports and that spent on imports.

Go Active

Find out the difference between a trade surplus and a trade deficit. Then, explain the terms trade surplus and trade deficit to a member of your family, using several objects from the kitchen as your props.

The Essentials

Usually, MEDCs **export** valuable **manufactured** goods such as electronics, generators, and cars and **import** cheaper **primary** products such as sugar, tobacco, flowers, tea and coffee.

In LEDCs the opposite is true. This means that LEDCs earn little and they remain in poverty, the country is forced to borrow money to pay for its imports and the country goes into debt.

The price of primary products fluctuates on the world market. Prices are set in MEDCs and producers in LEDCs lose out when the price drops. Ghana for example gets around 75% of its money from cocoa and timber. LEDCs very dependent on the world trade system yet they have little control over how it operates.

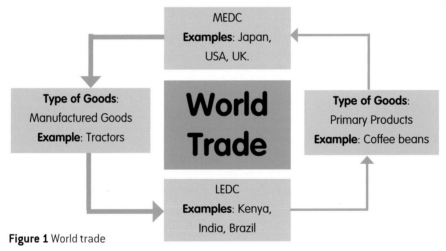

Figure 1 World trade

Increasing trade and reducing their balance of trade deficit is essential for the development of an LEDC. **Free trade** is the aim of many countries where governments neither restrict nor encourage trade. A disadvantage of this is that a country, such as the UK, could be swamped by cheap imports made in countries with lower labour costs. This is good for the consumer but could be bad for industry where many jobs may be lost. Some countries, particularly MEDCs, impose **tariffs** and **quotas** to protect themselves from the import of cheap products. Some pay a **subsidy** to its own farmers and businesses so that goods can be sold at cheaper prices to compete with imported goods.

GATT and the WTO

GATT (General Agreement on Trade and Tariffs) was founded in 1947 to encourage free trading. It took until 1993 before some agreements were signed. In 1995, GATT was replaced by the WTO (World Trade Organization). With 120 members it supervises the implementation of trade agreements, settles trade disputes and encourages free and fair trade.

Tariffs are taxes imposed on imports.

Quotas are limits on the amount of goods imported.

Subsidies are grants of money given by governments to maintain the price of a specific product e.g. milk.

41

Case Study – Kenya

Kenya is typical of an LEDC. A high proportion of it's people work in the primary sector; most are subsistence farmers. In recent years, however, tourism has become the country's main money earner. In 2008 Kenya imported goods to the value of $11,000 million and exported goods to the value of $5,000 million giving a massive trade deficit. 56% of Kenya's exports were agricultural products whereas 90% of its imports were fuel and manufactured goods.

Go Active

Trade quotas, **trade tariffs** and **subsidies** are all ways in which more powerful nations (normally MEDCs) control the trading of goods around the world. For each of these three trade controls, describe the effects they have on LEDCs. Remember to refer to Figure 1 for help. Try and consider the positive as well as the negative effects.

Inside Information

Understanding and using geographical terminology is vital for exam success. When reading a question you need to think about the marks available; this will guide the detail in your answer.

You need to think about the command terms which tell you what to do. Then you need to think about the key terms so you know what you are being asked to write about.

Question

Explain what is meant by a trading bloc.   [2]

Sample answers

Student A

A trading bloc is where countries can trade with each other without paying taxes. ✓

Student B

A group of countries✓ which have an agreement to export and import goods from each other without tariffs,✓ e.g. the European Union.✓

Examiner's comments

A straightforward question worth 2 marks. Every question and answer needs to be thought about carefully and every mark needs to be "fought" for in the game of exams. The command word in this question is "explain" and the key term is "trading bloc" which should be part of your geographical terminology. Student A scores only 1 mark. Student B makes three clear points, including giving the European Union as an example, and scores the maximum 2 marks.

EXAM SPOTLIGHT

1 Explain what is meant by globalisation.   [2]
2 Give 3 reasons to explain why there has been increased interdependence between MEDCs and LEDCs in recent years   [3]
3 What are the benefits of globalisation to newly industrialised countries such as China?   [6]

How are global patterns of development identified?

How is economic and social development measured and what are the global patterns?

More Economically Developed Countries (MEDCs): countries which have a high standard of living and a large GNI.

Less Economically Developed Countries (LEDCs): countries with a low standard of living and a small GNI.

The Essentials

Development

This describes the process of change which improves the well being of a society, in terms of material wealth and quality of life. Different countries of the world exhibit different levels of development:

- Economic development involves increased employment, income and usually industrial growth.
- Social development means having cleaner water, better standards of living, better access to education, better health, housing and leisure.
- Environmental development involves improving or restoring natural environments.
- Political development involves developing stable and representative governments.

Measuring development

In order to study development, geographers must first measure how developed one country is either compared to other countries or to the same country in the past. To measure development, geographers use a number of **indicators**.

Economic development indicators	Human development indicators
Gross National Income (GNI) previously known as Gross National Product (GNP) – the total value of all goods and services produced in a country in one year plus income from people living abroad.	**Life expectancy** – the average age to which a person lives.
GNI per capita – a country's GNI divided by its population.	**Infant mortality rate** – the number of babies per 1000 live births who die under the age of one year.
Gross Domestic Income (GDI) previously known as Gross Domestic Product (GDP) – the total value of all goods and services produced in a country in one year.	**People per doctor** – the number of doctors per 10,000 people.
Unemployment – measured by the number of people who cannot find work.	**Risk of disease** – the percentage of people with dangerous diseases such as AIDS, malaria and tuberculosis.
	Access to education – how many people attend schools and universities.
	Literacy rate – the percentage of adults who can read and write.
	Human Development Index (HDI) – a mix of indices that show life expectancy, adult literacy, education and GNP per capita.

Global patterns of development

Newly Industrialised Countries (NICs): countries that have, relatively recently, seen massive growth in their manufacturing industries. These countries, including South Korea and Taiwan, have been described as 'tiger economies', although they have seen less success in very recent years.

Recently Industrialised Countries (RICs): a term used for the very recent growth of India and China.

The Essentials

The Brandt Line

In 1980 the Brandt Report, written by German Chancellor Willy Brandt, divided the world into the rich North and poor South. However, the picture has changed considerably since, with many countries such as Brazil, India and China developing rapidly. Some even argue that the terms MEDC and LEDC are no longer appropriate to describe countries in various stages of development. The World Bank classifies the world's countries into four categories of wealth: Low Income, Lower Middle Income, Upper Middle Income and High Income.

Country	GNI per capita $US (2008)	Birth rate per 1000 (2008)	Death rate per 1000 (2008)	Life expectancy (2008)	People per doctor	Literacy rate % (2008)
Japan	38,210	8	9	83	613	100
UK	45,390	12	10	79	623	100
Brazil	7350	16	6	72	729	91
India	1070	23	8	64	2440	66
Kenya	770	39	12	54	10,000	74
Ethiopia	280	38	12	55	38,000	36

Here are six countries and six categories of information about each country. Examiners will assess your ability to study information, i.e. quote facts and figures that are presented to you and compare the same categories for different countries.

Below are five sentences, some of which include the two skills looked for by examiners ... and some which don't! Highlight the sentences you think are successful and improve the sentences you think are not.

a) Japan has a lower birth rate (8/1000) than the UK (12/1000).
b) Brazil is a poor country because its GNI per capita is only $7,350.
c) Ethiopia is better off than Kenya because of its number of doctors.
d) The literacy rate in India is much lower than that of Japan.
e) Life expectancy in Kenya is the lowest of the six countries.

What are the regional patterns of economic and/or social development in one LEDC?

Case Study – Ghana

- Ghana is a country in sub-Saharan Africa.
- The GNI is $770 per person (2008).
- 45 per cent of the population of Ghana live on less than $1 a day.
- 19 per cent of children are malnourished.
- Ghana suffers from a sharp north–south divide.
- The **south** has a long wet season, so farmers grow sorghum and cocoa.
- The **north** has unreliable rainfall, so farmers grow fewer crops and many keep goats.
- More people live in urban areas in the south.
- Most people work on farms, many as landless labourers.
- Incomes in urban areas are up to 2.5 times higher than those in rural areas due to some manufacturing and a growing tourist industry.

What progress is being made towards achieving the Millennium Development Goals?

What are the Millennium Development Goals (MDGs) and how are governments and non-governmental organisations addressing them?

The Essentials

In 2000 the United Nations set eight targets, known as the Millennium Development Goals, which aimed to promote human development.

> **Goal 1**: Eradicate extreme poverty and hunger
>
> **Goal 2**: Achieve universal primary education
>
> **Goal 3**: Promote gender equality and empower women
>
> **Goal 4**: Reduce child mortality
>
> **Goal 5**: Improve maternal health
>
> **Goal 6**: Combat HIV/AIDS, malaria and other diseases
>
> **Goal 7**: Ensure environmental sustainability
>
> **Goal 8**: Develop a global partnership for development

Some countries have made good progress towards these goals, whilst others, particularly African countries, have made little. In 2009 the UN reported what progress had been made towards achieving the goals:

- Globally, the proportion of hungry people decreased from twenty per cent in 1990 to sixteen per cent in 2006.
- Poverty levels have fallen dramatically in east Asia but elsewhere, progress has been slow, e.g. in sub-Saharan Africa. Here the poverty rate remains above 50 per cent.
- Global enrolment in primary education reached 88 per cent in 2007, up from 83 per cent in 2000.
- Deaths in children under five have fallen to nine million in 2007, down from twelve million in 1990. Child mortality rates remain high in sub-Saharan Africa, although the distribution of bed nets has reduced the number of deaths from malaria.
- The number of AIDS deaths peaked in 2005 at 2.2 million and has since declined. However, the number of people living with HIV worldwide – estimated at 33 million in 2007 – continues to grow.

Aid

Aid is the transfer of resources from richer countries to poorer ones. It includes money, equipment, training and loans. There are different types of aid:

- **Bilateral aid** is an arrangement between two countries. It is often 'tied aid', meaning that the receiving country has to spend the money on goods and services from the donor country.
- **Multilateral aid** is money donated by richer countries via organisations such as the International Monetary Fund (IMF), the United Nations (UN) and the World Bank.
- Short term **emergency aid** provides immediate relief during or after a disaster such as famine or a tsunami. It includes food, medicines and tents.
- **Long term aid**, such as education for young people, is a sustained programme which aims to improve standards of living.
- **Debt abolition** is when richer countries cancel debt owed to them by poorer countries.
- **Non-governmental aid** is given through charities such as Oxfam and Save the Children. This type of aid may provide emergency relief or may support small scale developmental projects, such as the building of wells to provide clean water.

Go Active

In favour of giving aid	Against giving aid
A Emergency aid in times of disaster saves lives and reduces misery.	F Aid can increase dependency on the donor country.
B The provision of clean water can lead to long term improvements in standards of living.	G Dependency on food aid slows improvements in agriculture.
C Developing natural resources and power supplies benefits the country's economy. Industrial development creates jobs.	H Profits from large projects can go to multinationals and donor countries.
	I Aid doesn't always reach the people who need it and can be kept by corrupt officials.
D Aid for agriculture helps increase food supplies.	J Aid can be spent on 'prestige projects' or in urban areas rather than in areas of real need.
E Medical training improves the standard of health.	K Aid can be used as a lever to exert political pressure on the receiving country.

1 Consider all the reasons for and against giving aid. Now draw a set of weighing scales and place the reasons for giving aid on one side and the reasons against on the other.

2 Then, give each reason a score out of three to reflect its value, i.e. if you think a reason is

superb, you would give it 3. If you think it's OK, you would give it 1.

3 Once you've done this, total up the score on each side of the scales to see if the reasons for or against aid are stronger.

4 List some advantages for a donor country giving aid.

What progress is being made by south Asian countries towards the MDGs?

UN Millennium Development Goals Report 2009 – south Asia

Progress in southern Asia has been slower than most other regions of the world with only a three per cent drop in extreme poverty rates, from 42 to 39 per cent. There are bright spots in the UN report, such as an eleven per cent gain in primary school enrolment.

The region is second only to sub-Saharan Africa in the proportion of people who are undernourished (21 per cent in 2008).

Southern Asia has achieved its target of cutting in half the proportion of people without access to clean water, but it is lagging behind in

providing access to safe sanitation with 580 million people still without sanitation.

Despite some gains for girls' school enrolment, southern Asian women remain at a huge disadvantage. Only nineteen per cent of paid jobs in the region are held by women.

Maternal health conditions also remain dismal. In 2009 the death rate of children born to South Asian mothers was 7–11 per cent.

Adapted from the United Nations Millennium Development Goals report, 2009.

Go Active

1 Read quickly the above UN Millennium Development Goals Report for south Asia. Once you've read the article, cover it up and decide which of the statements below are true:

a) The action against extreme poverty in southern Asia has been slower than in most other regions of the world.

b) There are no bright spots in the UN report.

c) Southern Asia has more under-nourished people than sub-Saharan Africa.

d) Southern Asia has achieved its MDG target of cutting in half the proportion of people in 1990 without access to water.

e) It is lagging behind in providing access to safe sanitation with 580 million people still without access.

f) Southern Asia recorded only a three per cent point drop in extreme poverty rates, from 42 to 39 per cent.

g) Despite some gains for boys' school enrolment, southern Asian men remain at a huge disadvantage.

h) Only nineteen per cent of paid jobs in the region are held by women.

2 Now re-read the article more carefully and repeat the exercise above. Don't forget to cover up the article so it can't be referred to!

The Essentials
Large scale development projects

Many LEDCs use large scale development projects to 'kick start' their economies and the development process. This top down development relies on the multiplier taking the benefits of the investment through the country. Examples of such projects are:

- the building of a dam to provide power and water, e.g. Narmada River Project, India
- the construction of roads such as the trans-Amazonian highway.

Large scale development projects rely heavily on foreign investment and aid. This money often comes from international organisations such as the World Bank.

They may also create a lot of problems. Money often needs to be borrowed from MEDCs which places the country in debt and large projects often cause environmental problems and disrupt local communities.

> The **multiplier effect** describes how an increase in economic activity starts a chain reaction that generates more activity than the original increase. For example, the setting up of a factory will give employees an income, this will lead to increased spending on consumer goods and new industries will develop to meet this demand. There will be a need to transport goods, so lorry drivers will need to be trained and schools will be set up to teach drivers and so on …

Case Study – The Narmada Development Project, India

- The Narmada project was first envisaged in the 1940s by the country's first prime minister, Jawaharlal Nehru.
- The multi-million dollar project involved the construction of some 3200 small, medium and large dams on the Narmada river. The aim was to provide the large amounts of water and electricity that are desperately required for development.
- The Sardar Sarovar is the biggest dam on the river and its construction has been fiercely opposed. The Narmada Bachao Andolan (Save the Narmada Movement) argue that the project will displace more than 200,000 people and damage the fragile ecology of the region. They claim it will drown important religious and cultural sites, submerge forest farmland, disrupt downstream fisheries and salinate land, increasing the prospect of insect-borne diseases. Some scientists have added to the debate, saying the construction of large dams could cause earthquakes.

- Those in favour of the project say that the project will supply drinking water to 30 million people and irrigate crops to feed another 20 million people. Water for irrigation would allow marginal land to be used for agriculture and relatively cheap electricity (from HEP) will help India along the path of development. Many people in India live in poverty and those in favour claim that this project will lead to industrial growth, wealth creation and will alleviate the suffering of millions.
- The World Bank originally sanctioned a loan of $450 million for the project but then withdrew support, citing concerns about resettlement programmes and environmental damage. In 2000 the Indian Supreme Court gave the final go-ahead for the construction of the Sardar Sarovar dam. $200 million was borrowed for the project to go ahead.
- 'Big Dams are to a nation's "development" what nuclear bombs are to its military arsenal. They're both weapons of mass destruction.' Arundhati Roy, a NBA supporter.

Go Active

Create a case study card for the Narmada River project.

What progress is being made by sub-Saharan African countries towards the MDGs?

UN Millennium Development Goals Report 2009 – sub-Saharan Africa

The proportion of hungry people was driven down from 32 per cent in the early 1990s to about 28 per cent today. However, the current rate of progress is not sufficient to meet the target set for 2015.

The proportion of working poor, those who are employed but do not earn enough to raise themselves and their families above the poverty line, is 64 per cent of the working population.

African women as mothers face the highest mortality rate among all developing regions. Half of the world's maternal deaths occur here and more than half of all births occur without the assistance of trained personnel.

By 2007, 73 per cent of children were immunised against measles, up from 55 per cent in 2000. Insecticide-treated anti-malaria bed nets are reaching African children more widely, with their use up from two per cent in 2000 to twenty per cent in 2006. However, in 2007 nearly one in seven children died before their fifth birthday.

The prevalence of HIV has levelled off and access to treatment has improved but sub-Saharan Africa is still home to 67 per cent of people living with HIV. Over one third of new HIV infections and 38 per cent of AIDS deaths in 2007 occurred in the continent.

Almost half of the 72 million children out of school worldwide live in sub-Saharan Africa. This number is dropping far too slowly for the education target to be reached by 2015.

Sanitation coverage has improved markedly, with the proportion of people using toilets, latrines and other forms of improved sanitation increasing by over 80 per cent since 1990.

The total amount of aid given to the region remains well below the United Nations target of 0.7 per cent of donor country gross national income.

Adapted from the United Nations
Millennium Development Goals report, 2009.

Go Active

Draw a collection of diagrams to represent at least five pieces of information you think are important from the report above. You are not allowed to use any words in your diagrams. The images must speak for themselves but you can use figures.

The Essentials

Small scale local development

Small scale development projects rely on making small changes, working with local people and using local skills. This can be described as bottom up development. Small scale projects aim to meet people's everyday needs, such as clean water and sanitation. This development does not rely on heavy investment, as it is often supported by non-governmental organisations (NGOs), and uses appropriate technology.

Case Study – Water, sanitation and hygiene education in Ghana

Access to water and sanitation facilities in Ghana is low, particularly in rural areas. Only 50 per cent of the rural population in Ghana has access to clean water.

The main sources of water in many parts of rural Ghana are small ponds and unprotected wells, both of which are easily polluted, causing disease and ill-health. Oxfam is working with WaterAid, UK and a local partner, Rural Aid to provide hand dug wells with pumps and constructing ventilated pit latrines.

Nyama Akparibo, aged 28 lives in Asamponbisi Village, in Eastern Ghana. WaterAid and Rural Aid helped to build a well in her village.

'I have lived here for 15 years. I have three children, aged ten, eight and six. The water pump was put in three years ago. Our community helped the "water people" (Rural Aid) build the well. The men did the digging and we women collected sand and stones, helped clear away the sand when the well was dug and helped carry the mortar for construction. We needed a well because we used to drink bad water and get sick, especially the children. We got diarrhoea and stomach pains and felt very weak. We used to be so weak we couldn't do any work. Our main work here is basket weaving, but when you have diarrhoea you are too weak to sit and weave.

'I used to collect water from a stream over there. Animals used to drink from there too and the water was not very clear. When I had

my first baby I had to use water from the stream and the baby and I were always sick. Now my children no longer get diarrhoea and we have enough water to bath our children before they go to school, prepare food for them, wash and do our chores. It used to be very time consuming to collect water at the stream as others were already there and you had to queue while they filled their pots before you could collect your own. Now we can just come to the pump and there is always plenty of water.'

http://www.oxfam.org/en/development/ghana/
hygiene-education © Oxfam International

Go Active

Look at the Case Study above. 'Small Is Beautiful: Economics As If People Mattered' was an essay written by British economist E.F. Schumacher in 1973. It is often used to champion small development projects that are believed to empower people, in contrast with the belief that 'bigger is better'.

Do you agree that 'small is beautiful'? Imagine you were in charge of a television programme in which this topic was to be debated. What arguments could be put forward by the audience to both support and argue against Schumacher's ideas?

Inside Information

a) Describe and suggest possible reasons for the relationship between the Infant Mortality Rate and GDI per capita as shown in Figure 1. [4]

Sample answer

The graph shows that the lower the GDP per capita, the higher the infant mortality rate✓, e.g. Sierra Leone✓ has a very high mortality rate but a very low GDP. This is probably because if a country is poor, it will not have enough money for health care✓ or nutrition✓ to stay healthy.

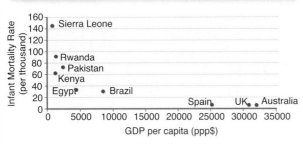

Figure 1 Scattergraph of IMR and GDP per capita

Examiner's comments

This is a good answer. The candidate understands the command words, correctly describes the relationship and develops the point by giving Sierra Leone as an example, turning one mark into two. Candidates often score poorly when there are two command words in a question but this candidate goes on to give two clear points to explain the relationship. The candidate clearly understands the terms 'infant mortality' and 'GDP' and I am impressed by the term 'nutrition'.

- It is important that you have a good geographical vocabulary and that you can use it well. If you don't understand geographical terms, you may not understand the demands of a question. If you have a good geographical vocabulary, you will impress the examiner with your answers.
- A good way of building up your geographical vocabulary is to compile a card index of terms associated with each theme of the exam.

The Human Development Index (HDI): uses life expectancy, literacy, years in education and income per person to measure development on a scale of 0 to 1.

EXAM SPOTLIGHT

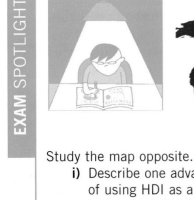

Study the map opposite.

i) Describe one advantage of using HDI as a measure of development. [2]

ii) Describe the distribution of countries with HDIs of less than 0.6. [4]

iii) Explain how long term development aid can help improve the quality of life of people in countries with a low HDI. [6]

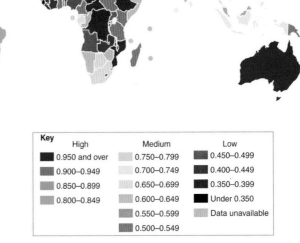

Key		
High	Medium	Low
0.950 and over	0.750–0.799	0.450–0.499
0.900–0.949	0.700–0.749	0.400–0.449
0.850–0.899	0.650–0.699	0.350–0.399
0.800–0.849	0.600–0.649	Under 0.350
	0.550–0.599	Data unavailable
	0.500–0.549	

Figure 2 The Human Development Index (HDI)

What are the coastal processes and what landforms do they create?

What processes are associated with the sea?

The Essentials

- Waves are created when wind blows over the surface of the sea.
- **Destructive waves** are formed by strong winds and erode the coastline.
- **Constructive waves** are gentle waves which deposit beach material.
- The **fetch** of the wave is the distance the wind has travelled over the sea before reaching the coast.
- Waves **break** when they reach shallow water.
- Water that moves up the beach is called the **swash** and water which moves down the beach is known as the **backwash**.

Erosion	Transport	Deposition
Hydraulic action: the force of waves crashing into cliffs. Air trapped in the cracks is compressed, which breaks up the rock. **Abrasion** (some use the term **corrasion**): waves hurl sand and pebbles against the cliff, which wears the land away. **Solution** (some use the term **corrosion**): salt water dissolves rocks made of calcium carbonate, e.g. limestone. **Attrition**: pebbles are rolled back and forth. They collide together, which makes them smaller and more round, eventually turning them into sand.	Prevailing winds blow at an angle to the coastline. When a wave breaks, the swash carries mud and sand up the beach at an angle. Gravity pulls sediment directly down the beach in the backwash. Each wave repeats this action and material is moved along the coastline in a process known as **longshore drift**.	Beaches of sand and shingle are formed when constructive waves lack sufficient energy to transport material.

Go Active

1 Use the 'cover and reveal' method for each of the four types of erosion, i.e. read the information, cover what you have read, then try to write it word for word. Repeat this activity until you are happy with what you've remembered.

2 Gather together some objects such as string or dried spaghetti, a pebble, a piece of paper and a pen. Now use these materials to explain the transportation process of longshore drift to a friend. Avoid using paper and pen as much as you can. The object of this activity is to *show* movement, rather than just writing about it.

What landforms result from these processes?

The Essentials

Cliffs and wave cut platforms

- A cliff is a steep rock face.
- Waves erode or 'cut' a notch at the base of the cliff, between high and low tide.
- Cliffs eventually collapse, having been undercut, and the cliff face moves inland.
- Over many years the process is repeated. The cliff retreat leaves a long rocky platform known as a **wave cut platform**.

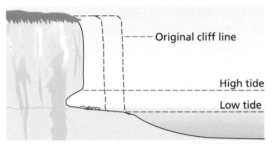

Figure 1 Cliff retreat

Go Active

Annotate Figure 1 to explain the formation of a wave cut platform.

Headlands and bays

- A **headland** is an area of land that juts into the sea and is formed out of harder, more resistant rock.
- A **bay** is formed between the headlands. Beaches often form in more sheltered bays.

Caves, arches, stacks and stumps

These are the landforms of erosion:

- Natural **faults** in the headland are eroded by the sea.
- Faults develop into cracks and then into a **cave**.
- Caves often form on both sides of the headland and break through to form an **arch**.
- The arch is weathered and eroded until it collapses to leave an isolated piece of rock known as a **stack**.
- The stack is eroded at its base to form a **stump**.

Beaches, spits and bars

- A **beach** is a build up of sand, shingle and pebbles deposited by constructive waves.
- Longshore drift transports beach material along the coast. Where the coast changes direction, e.g. at a river mouth, beach material is carried out to sea. This creates a new strip of land called a **spit**.
- The end of the spit will often be hooked if wind sometimes blows from a different direction. Silt is then deposited in the sheltered water and a **salt marsh** is formed, such as Spurn Head in Humberside.
- A **bar** is similar to a spit but forms across a bay from headland to headland, such as the sand bar at Slapton Ley, Devon.
- A **tombolo** is also similar to a spit but links the mainland to an offshore island, e.g. Chesil Beach in Dorset.

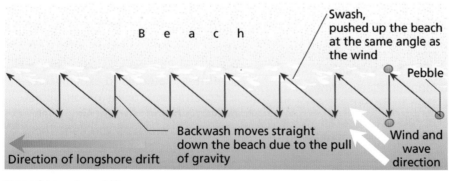

Figure 2 Longshore drift

Weathering: the break up or decomposition of rock 'in situ'. There are three types of weathering: **physical** weathering including 'freeze thaw' action, **chemical** weathering which forms cave systems in limestone rocks and **biological** weathering.

Case Study – The Dorset Coast

Dorset is famous for the Jurassic Coast World Heritage Site, spectacular coastal scenery and the holiday resorts of Bournemouth and Weymouth. Jutting out into the English Channel is a limestone island called the Isle of Portland. This is connected to the mainland by Chesil Beach, a tombolo.

Other landforms along the Dorset coast include:
• Resistant limestone and chalk headlands.
• Bays formed in the softer clay.
• Beaches in the sheltered bays.
• Stacks such as 'Old Harry'.

• Lulworth Cove formed where the sea has broken through the hard limestone.
• Arches such as Durdle Door.

Figure 3 Durdle Door

Go Active

1 Without re-reading the information in this section, try to list all the coastal processes and landforms you've just revised. Once you've done this, look back at the information on the previous pages and see how many you remembered and listed correctly.
2 Now draw a labelled diagram for each of the listed coastal features.
3 Empty out the contents of your pencil case and select at least five objects.

Use these objects to explain each of the coastal processes in this section. Remember, by trying to 'make' something in this way, you are having to consciously process your thinking. This, in turn, improves your memory and understanding of the information.

4 Create a revision card (remember those postcard sized pieces of card?) to help you revise the Dorset Coast case study. Don't forget to include a location map, keywords and diagrams.

How do these landforms and processes affect the lives of people living along the coast?

The Essentials

Over 17 million people live within 10 km of the UK coast and 40 per cent of the country's manufacturing industry is on or near the coast. Coastal processes and their resultant landforms will therefore impact on large numbers of people and there are many different groups who have an interest in what happens in coastal areas including:

• Residents
• Environmental groups
• Industrialists
• Local councils

• National governments
• Tourist boards
• National park authorities, such as the Pembrokeshire National Park Authority

Issues in coastal areas include:
• Erosion threatens some coastal settlements.
• New tourist attractions proposed.
• Existing tourist resorts are in decline.
• People want to build new housing in the attractive coastal environment.
• Danger of flooding if sea levels rise.
• Problem with sewage and/or pollution.

Each interest group may have a different view about what should be done to protect and manage coastal areas.

Go Active

Use a 'mind map' technique to summarise how coastal landforms and processes can affect the lives of people.

How are coasts managed?
What are the advantages and disadvantages of hard and soft engineering strategies?

The Essentials

Coastal environments need to be managed to ensure:
- human activities are protected from erosion and flooding
- important habitats and heritage sites are preserved.

Hard engineering	Soft engineering
This uses structures or machinery to control coastal processes. They tend to be expensive, short term options and have a high impact on the landscape or environment. They also tend to be unsustainable. Some examples include: **Sea walls**: concrete walls designed to reflect the energy of waves and prevent flooding. **Groynes**: wooden (usually) barriers, built down a beach, trap sand being transported by longshore drift. The resultant, wider beach absorbs wave energy. **Rip rap**: large boulders of hard rock absorb the energy of the waves. **Gabions**: steel mesh cages, containing boulders, absorb the energy of the waves.	This involves working with nature. It is often less expensive and usually provides more long term, sustainable solutions that have less of an impact on the environment. Some examples include: **Beach nourishment**: beach material is replaced by sand from further along the coast or dredged from the sea bed. **Managed retreat**: areas of the coast are allowed to erode and flood naturally. This will usually be areas considered to be of low value.

Go Active

1 Below are some of the advantages and disadvantages of hard and soft engineering.

2 Complete the Venn diagram below by placing the corresponding numbers in the correct place. Each number can only be placed once.

3 Once you have completed this diagram, decide whether hard or soft engineering is preferable. Remember that not all advantages and disadvantages will be equally weighted. The fact that there are more advantages than disadvantages to hard engineering doesn't make it better than soft engineering, because the disadvantages may be of greater importance.

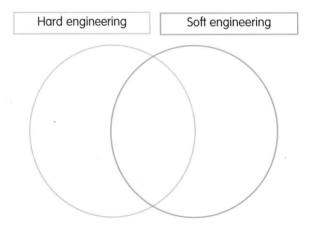

Hard engineering   Soft engineering

Advantages	Disadvantages
1 They are very effective.	8 They can be ugly.
2 Some have little impact on the rest of the coastline.	9 They are generally very expensive.
3 Some are quick to install.	10 Some are newer approaches and little is known about their success.
4 There is generally no costly building involved.	11 Offshore dredging alters the direction of the waves and can lead to erosion further down the coast.
5 They work with the natural processes.	12 Some of the building materials and implantation can be expensive.
6 Some can last a long time.	13 Some can be easily eroded by the waves.
7 Some can be quite cheap.	

How should coastal environments be managed in the future?

Why are sea levels changing and how will these changes affect people?

The Essentials

Global warming is likely to cause a rise in sea level due to the melting of the ice caps. This will cause flooding in many low lying areas of the world.

Case Study – The Thames Gateway

The Thames Gateway project aims to build 160,000 affordable homes by 2016 across the south east of England. Many of the houses will be built on brownfield sites but will be in areas at risk from flooding. The project will also see the creation of 225,000 new jobs.

Planners have adopted a holistic approach to flood prevention. They say that this will replace the expensive and unsustainable 'high walls approach', where everything stays dry. Flood-prone areas will see precautions like a 'green bund' (earth embankment) placed between houses and the river. Tiered flood defences, which allow the river to rise in steps, will also be implemented.

The Environment Agency has said that planning allows the early placement of flood management. The Association of British Insurers has stated that sensible building depends upon building secure flood defences, while Friends of the Earth said it doubts that people would buy these houses when sea levels are rising due to global warming.

Go Active

1 Imagine you are the chairperson of a meeting where the Thames Gateway plan is discussed. At the meeting are representatives of builders, the government, Friends of the Earth, the Environment Agency, insurers and the public. Summarise the thoughts of each of the delegates and make a decision about whether you think the plan should go ahead. Justify your decision.

2 Highlight the three sentences from the list below that you believe are true:

   a) Latest estimates from the UK Climate Impacts Programme suggest that the east coast of England could see further rises in relative sea levels of 40 cm by 2050.

   b) Latest estimates from the UK Climate Impacts Programme suggest that the east coast of England could see further rises in relative sea levels of 100 cm by 2050.

   c) By 2080 the sea level on the east coast of England may have risen by 80 cm.

   d) The number of properties at risk of flooding in eastern England will rise by 48 per cent from 270,000 to 404,000 following a rise in sea levels of 40 cm.

   e) The number of properties at risk of flooding in eastern England will rise by 25 per cent from 270,000 to 404,000 following a rise in sea levels of 40 cm.

What is the most sustainable way to manage our coastline in the face of rising sea levels?

Many argue that we need a strategic approach to managing our coastline. Interested groups need to work together to make sensible, long term decisions about the way we use our coasts.

They will need to consider protecting coastline from future flooding and possibly relocating people and homes, or abandoning agricultural land.

Go Active

1 Below are three suggestions for the future management of our coastline. Consider points for and against each suggestion.

   a) Build a giant sea wall to protect the most vulnerable areas. The material for this sea wall should be transparent plastic bottles that have been recycled.

   b) Import sand from the Sahara in order to build giant sand dunes along the coastline.

   c) Retreat the line. This means punching a hole in an existing coastal defence to allow land to flood naturally between low and high tide (the intertidal zone).

2 How do you think our coastline should be managed in the future?

Case Study – The Welsh coastline

BBC News – Action call on disappearing coast

Tuesday, 13 February 2007

Wales has been urged to take 'urgent action' to prepare for the impact of coastal erosion and flooding.

The National Trust says three-quarters of the Welsh coastline it owns could be badly affected over the next century. By that time, some experts predict that sea levels will rise by a metre and climate change will lead to more severe storms.

The NT said it wanted a fresh approach, with 'urgent action to put in place coherent, long-term planning to address the massive impacts of future sea level rise'.

'No-one in Wales is more than 50 miles from the shore, and tourism is particularly dependent on our wonderful coast.

'Like King Canute, we can't control the ocean and command it to retreat. Instead, we must plan how to adapt to a future of advancing seas. The first step is to raise awareness of what is at stake. This should strengthen the call to reduce our carbon footprint, but we also need to adapt to the changes underway and plan for the future of coastal communities.'

'If we are to succeed then we will need to think innovatively in terms of how we manage our coastline and ensure that we engage the whole of society in the process.'

BBC News website: http://news.bbc.co.uk/1/hi/wales/6356073.stm

---

Go Active

1 Quickly read the BBC news report above and decide which of the statements below are true. Cover up the article once you've read it so that you can't refer to it when selecting the 'truths'.

   a) One quarter of the Welsh coastline could be badly affected over the next century.

   b) Sea levels will rise by a metre and climate change will lead to more severe storms.

   c) No one in Wales is more than five miles from the shore.

   d) We must plan how to adapt to a future of retreating seas.

   e) We can control the ocean and command it to retreat.

   f) We must ensure that we engage the whole of society in the process.

   g) We need to think innovatively in terms of how we manage our coastline.

2 Re-read the article more carefully and repeat the exercise above. Don't forget to cover up the article so it can't be referred to.

TOP SECRET

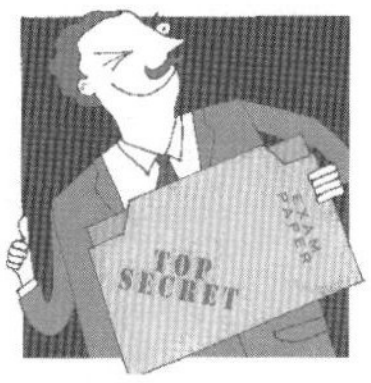

Inside Information

Explain the formation of bays and headlands. You may add to the diagram below to help your answer. [6]

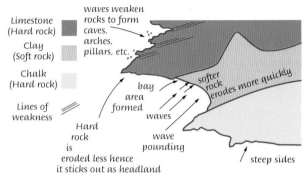

Limestone (Hard rock)
Clay (Soft rock)
Chalk (Hard rock)
Lines of weakness
waves weaken rocks to form caves, arches, pillars, etc.
softer rock erodes more quickly
bay area formed
waves
Hard rock is eroded less hence it sticks out as headland
wave pounding
steep sides
As the waves pound the coastline they erode the rock via hydrálic action, corrosion and wave pounding. The soft rock is eroded quicker than the hard rock hence the soft clay recedes ~~faster~~ further back than the hard rock. The headlands are formed as a result of the bay formation, they are areas of hard rock which stick out into the sea.

Figure 4 Student answer

Examiner's comments

The command word in this question is **explain** and the key words are **headlands** and **bays**. Although the question only states that you *may* add to the diagram, it would certainly impress the examiner and persuade him/her to give a high mark if you did. This answer is a good answer. The candidate clearly understands the formation of bays and headlands. Coastal processes are identified and the candidate has added to the diagram. However, the candidate has not given detail on how the processes of erosion operate, has mis-spelt 'hydraulic' and has incorrectly used the term 'wave pounding'. Therefore, although this is clearly a Level 3 answer, it is not worth 6 marks. I would award it 5.

EXAM SPOTLIGHT

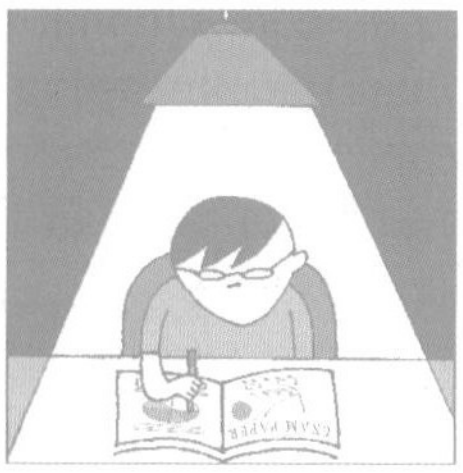

a) Study the Ordnance Survey map extract and draw an outline sketch of the area.

i) Mark with an arrow and label four landforms resulting from coastal erosion. [3]

ii) Explain the formation of a stack. You may draw a diagram if it will help your answer. [6]

b) Explain the difference between hard and soft engineering approaches to coastal defence. [3]

c) Give the advantages and disadvantages of any one method of coastal defence. [4]

Middlecombe Fm
Tinsey Head
Batton
Kellaton
South West Coast Path
Muckwell
Greenstraight
Hallsands
Bickerton
Hollowcombe Head
Hare Stone
Down Fm
Higher Borough
Shoelodge Reef
Start Fm
Nestley Point
Lannacombe Beach
The Narrows
START PC
Two Stones
Ravens Cove
Black Stone
Ballsaddle Rock
Sleaden Rocks
Lannacombe Bay
Gorah Rocks
40
39
38
37
36
35
79
80
81
82
83
84

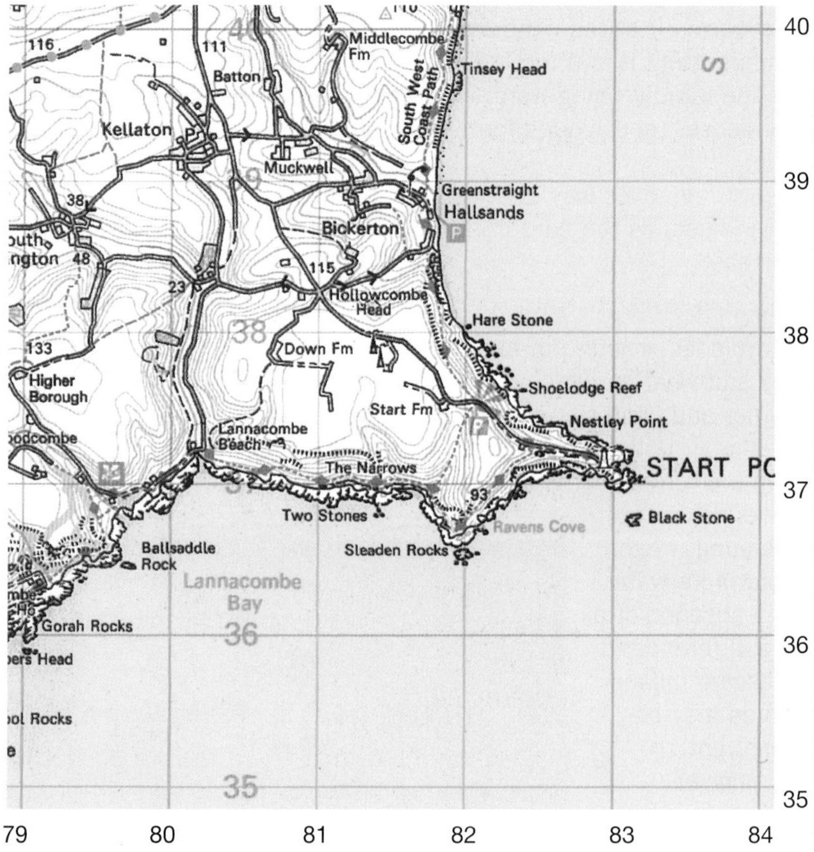

Figure 5 1:50,000 OS map of Start Point, South Devon (Landranger 202)

What are the differences in climate within the UK?

What factors create the variations in weather and climate experienced within and around the British Isles?

The Essentials

Britain has changeable weather because there are many factors which influence daily atmospheric conditions. The most important are low and high pressure systems. Wind direction, height of the land and time of the year are also vitally important.

> **Weather**: the atmospheric conditions at a particular place and time together with the changes taking place over the short term. Weather conditions include temperature, precipitation, wind, cloud and sunshine.

Low pressure systems

Depressions produce cloudy, rainy and windy weather. These low pressure systems often begin in the Atlantic and move east towards and across the British Isles. They are responsible for very changeable weather and frontal rainfall.

- At the warm front, lighter, warmer air from the south meets cooler air from the north and rises over it. As the warm air slowly rises, it cools and its water content condenses, forming clouds. This produces steady rain.
- Behind the warm front is an area of warm, rising air and low pressure. As this part of the depression passes over, there may be a short period of clear, dry weather.
- At the cold front, heavier, cooler air meets the warm air at the centre of the depression, undercutting it and forcing it steeply upwards. As the rapidly rising warm air cools, its water condenses, and clouds form. The result is heavy rain.
- Heavy rain gives way to showers and finally to clear skies, as the cold front moves away to the east.

High pressure systems

Anticyclones or high pressure systems give clear skies and weather which differs significantly in summer and winter.

- As the air doesn't rise but sinks, no clouds or rain are formed. This is because, as the air sinks, it warms meaning it can hold more water.
- The absence of a big difference in pressure means winds may be very light.
- In **summer**, anticyclones bring dry, hot weather. In winter, clear skies may bring cold nights and frost.
- Winter anticyclones may also bring fog and mist as the cold forces moisture in the air to condense at low altitudes.
- Summer anticyclones and their hot conditions may cause bubbles of air to rise rapidly, which leads to convectional rainfall and thunderstorms.

Rainfall

Precipitation is a term used to describe all types of moisture that fall from the sky. It happens when air rises. This can happen in three main ways, although in Britain we experience more **frontal** and **relief** rainfall than **convectional**.

- Relief rainfall occurs when the prevailing winds blow from the south west.
- Warm, moist air rises over mountains in the west of the British Isles.
- The air cools and water vapour condenses.
- Clouds form and rain falls over the mountains.
- To the east of the mountains, air descends, warms up and clouds evaporate.
- This area is called the rain shadow, where rainfall totals are much lower.
- Convectional rainfall occurs in summer when the sun heats the ground. Air next to the ground becomes hot and rises. This rising air cools, condenses and often gives heavy rainfall and sometimes thunderstorms.

Figure 1 Diagram of a depression

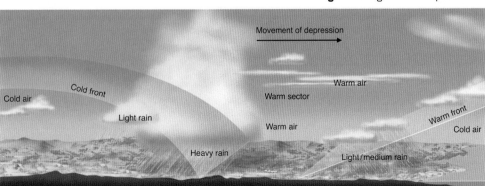

Go Active

Explain the relief rainfall shown in Figure 2 by completing the table.

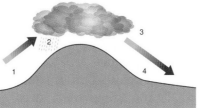

Figure 2 Relief rainfall

No.	Explanation
1	
2	
3	
4	

The climate of the British Isles

Britain has a **temperate maritime climate**, heavily influenced by latitude and the sea. The prevailing wind direction in Britain is from the south west. This brings moisture and rainfall to the west of Britain from the Atlantic Ocean. Mountains in the west intensify this rainfall.

There are five main factors that affect the climate in Britain:

- **Latitude** – Britain experiences a hot season (summer) and a cold season (winter), depending on whether the northern or southern hemisphere is tilted towards the sun.
- **Altitude** – Temperatures decrease by 0.6 °C for every 100 metres of elevation.
- **Continentality** – The sea heats up and cools down more slowly than the land. Therefore, coastal areas have a smaller temperature range than those inland.
- **Ocean currents** – The North Atlantic Drift brings warm water to our shores and keeps our climate mild in winter and cool in summer. The effect is particularly noticeable in the west when south westerly winds are blowing.
- **Wind direction** – The term 'prevailing wind' refers to the most common wind in an area. A wind will bring warm or cold, wet or dry weather depending on its source.

Air masses

- North westerly wind brings **polar maritime** air, giving cool and showery weather.
- South westerly wind brings **tropical maritime** air, giving mild and rainy weather.
- South easterly wind brings **tropical continental** air, giving hot and dry weather.
- Easterly wind brings **polar continental** air, giving hot weather in summer and cold weather in winter.
- Northerly wind brings **Arctic** air, giving cold weather with snow in winter.

> **Air masses**: are very large volumes of air with uniform temperature and humidity.

How does the weather create hazards for people?

What are the weather hazards associated with high and low pressure systems over the British Isles and with tropical storms?

The Essentials

Low pressure

When a series of areas of low pressure affects the British Isles the rainfall total can build and result in serious flooding. Low pressure areas are also associated with strong winds.

Go Active

Complete a mind map to show different ways in which the weather affects the lives of people.

Case Study – Gloucester floods

The people of Gloucester awoke on Friday 20 July 2007 to heavy rain and warnings of more torrential downpours to come. This deluge would make history by triggering Gloucestershire's biggest ever peacetime emergency. The relentless rain left hundreds of homes and businesses under water and thousands of drivers stranded, as road and rail routes were inundated by flood waters.

Soon the taps ran dry for half the county's population as a water treatment station on the outskirts of Tewkesbury was overwhelmed by the floods. Emergency services faced a desperate battle to save the county's power supplies as the Walham electricity sub-station was threatened by the rising flood waters.

High pressure

High pressure areas are anticyclones and are associated with ice and fog in winter and drought, if prolonged, in summer months.

Case Study – The drought of 2004

The years 2004–2006 were one of the driest periods on record in the UK. The south east of England is particularly vulnerable to drought due to the high population density – 13 million people live in the region – and the demand for water resources. There are also few reservoirs, so there is a heavy reliance on groundwater supplies. Two consecutive dry winters meant that these supplies were not replenished.

- Hosepipe bans were introduced in an effort to conserve water.
- Ground water fell to its lowest level on record.
- Some rivers dried up.
- Low river flow means that pollution will have a greater impact on the environment.
- Fish are more likely to die because of low flow, low oxygen levels, and higher water temperatures.
- Kew Gardens introduced a range of measures to conserve water, e.g. only watering newly planted trees, and newly turfed areas.

Tropical Storms

Tropical storms are called **hurricanes**, **typhoons**, **cyclones or willy-willies** – different names are used in different parts of the world. If these huge storms start in the Atlantic and move west, they are called hurricanes.

- Hurricanes form when sea temperatures are over 27 °C.
- In an average year, over a dozen hurricanes form over the Atlantic Ocean and head west towards the Caribbean, Central America and southern USA.
- Hurricanes may last as long as a month and wind speeds can reach over 120 km/h.
- The intense winds of tropical storms can destroy whole communities, buildings and communication networks.
- Heavy rainfall causes flooding.
- The winds generate abnormally high waves, and low pressure creates tidal surges, which cause flooding in coastal areas.
- Tropical storms die out when they reach land and are cut off from their power source – the sea.

Case Study – Cyclone Nargis: Burma (LEDC)

Cyclone Nargis hit the coast of Burma (Myanmar) in May 2008. Nargis formed over the warm waters of the Bay of Bengal. The low pressure of the storm allowed sea levels to rise 3.6 m, creating a storm surge. The storm hit the low lying, densely populated coastal plain at high tide and caused extensive flooding for long distances inland. Further damage was caused by strong winds, up to 215 km/hr, creating huge waves up to 7.6 m above the storm surge.

It is estimated that 800,000 homes were damaged in the storm. Refugee camps were set up to give shelter to over 260,000 people. Damage was estimated at over $10 billion (USD). The exact death toll is unknown but it is estimated that at least 138,000 people

died most from drowning. Rice crops were destroyed adding to general food shortages. Drinking water supplies were contaminated by sewage. By June 65 per cent of the survivors were reporting health problems such as fever and diarrhoea. Many health centres had been destroyed in the flood so there was limited access to medical treatment. Relief efforts were slowed as Burma's military rulers initially resisted aid.

It estimated that 200,000 families had rebuilt their own homes within a year of the disaster. But recovery has been slow and hampered by a lack of political co-operation e.g. out of US $150 million requested for shelter repair and reconstruction only US $50 million has so far been received.

Case Study – Hurricane Katrina, August 2005

Location: Louisiana, USA

Effects:

- Storm surges reached over six metres in height and breached flood defences.
- New Orleans is below sea level and protected by levees. Following Katrina, 80 per cent of the city was flooded.
- Despite an evacuation order, many of the poorest people remained in the city and over 1200 people died.
- People sought refuge in the Superdome stadium but conditions were unhygienic and there was a shortage of food and water.
- Looting was commonplace throughout the city.
- One million people were made homeless.
- Oil facilities were damaged and, as a result, petrol prices rose in the UK and USA.
- The cost of the damage was estimated as $80 billion.

How do weather hazards affect people, the economy and the environment?

Go Active

1 Discuss what you think would be the social, economic and environmental consequences of flooding on the village of Tewkesbury, Gloucester. What do you think could have been done to lessen the impact of the flooding? Think back to 'Theme 1: Water'.

2 Make a list of any measures that could be taken to prevent this happening again in the future.

Can we manage weather hazards?

How can technology be used to forecast extreme weather and reduce the impact of its effects?

The Essentials

Forecasting the weather

Weather information can be obtained from:

- daily readings of instruments in a weather station
- automatic instruments in remote locations, such as ocean buoys
- ships and aeroplanes
- satellite and radar images.

Reducing the impacts

Modern equipment can track hurricanes to tell people when and where they will strike. This is useful as:

- storm warnings can be issued
- people can be evacuated
- preparations can be made for the storm, such as storing food and water

- windows can be boarded up
- storm shelters can be built.

MEDCs have the resources and technology to predict and monitor the occurrence of weather hazards, e.g. using satellites and specially equipped aircraft to train the emergency services appropriately and to educate people about necessary precautions.

LEDCs are often less prepared and may rely on aid from MEDCs for the rescue and recovery process. In these countries, measures taken to reduce the impacts of tropical storms include:

- building shelters capable of housing many families
- strengthening and raising earth banks along the coast and rivers
- planting mangrove trees along the coast, allows silt to build up which will absorb storm waves
- educating people about the risk.

Inside Information

a) Explain how a warm front brings cloud and rain.
You must add to the diagram below to help your answer. [6]

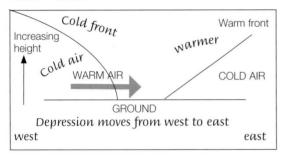

Depression moves from west to east

west   east

The warm front is a belt of cloud and continuous rain. In the cold front is a belt of shower clouds and heavy rain. When the warm and cold fronts join together it is usually a belt of rain.

Figure 3 Student answer

Examiner's comments

The command word in this question is **explain** and the key words are **warm front** and **rain**. Note also that the question states you must add to the diagram. It is likely that this question would be marked using a level marking scheme.

This is a weak answer. The candidate shows little understanding of the processes which lead to cloud and rain forming at the front. The text simply restates the information in the question that rainfall is associated with fronts. Although there is addition to the diagram, it does not explain how fronts give rainfall. I would give this answer 1 mark.

EXAM SPOTLIGHT

a) Re-write a Level 3 answer to the question in the 'Inside Information' box above. [6]
b) Explain why the mountainous parts of Wales have high amounts of rainfall. [5]
c) Study Figure 4. Describe and explain how altitude can influence UK rainfall totals. [6]
d) Explain how the impact of tropical storms can be reduced. [4]

Key
- Highland
- Low land

Key
Average annual rainfall
- Under 750 mm
- 750–1250 mm
- 1250–2000 mm
- Over 2000 mm

Figure 4 Altitude (left) and rainfall in the UK

What are biomes and how do they differ?

How does the physical environment interact with living things to produce different large scale ecosystems?

The Essentials

Large scale ecosystems

- The living parts of an ecosystem are called **biotic** and the non-living parts **abiotic**. The living and non-living parts of an ecosystem are interdependent, i.e. they depend on each other to make the system work. If one part of the system is altered, the whole system will be altered.
- Ecosystems can exist at a variety of scales, from a rock pool at the seaside to global systems known as **biomes**. Important processes within an ecosystem include the **nutrient cycles**, **energy flows** and **succession**.

The nutrient cycle

All living things need nutrients to live and grow. Nutrients are found in water, rocks and the atmosphere and they move through an ecosystem in a cycle, as follows:

- Weathered rock releases nutrients into the soil.
- Water is added to the soil by rainfall.
- Plants absorb the nutrients through their roots and leaves.
- Animals gain nutrients by eating plants.

> **Ecosystem**: the links between plants, animals and the non-living things around them, such as rocks, soil, water and climate.

- Plants and animals die and are decomposed by bacteria and fungi.
- Nutrients are returned to the soil.

Nutrients may be lost to the system by **leaching** and by being washed away in **surface runoff**.

The energy cycle

- Plants are the producers of energy in an ecosystem through photosynthesis, which uses energy from the sun.
- Animals (**herbivores**) eat the plants, **carnivores** eat the herbivores and so energy moves through the **food chain**.
- Food chains are connected to make a **food web**.
- The number of living organisms decreases at each stage of the food chain because energy is lost, used up in transpiration, movement and breathing.

Thus, the greater the inputs into an ecosystem, i.e. soil nutrients, water and sunlight, the greater the volume and diversity of plants and animals that can be supported. Arctic areas therefore have limited inputs of energy and, as a result, the number and variety of plants and animals are limited.

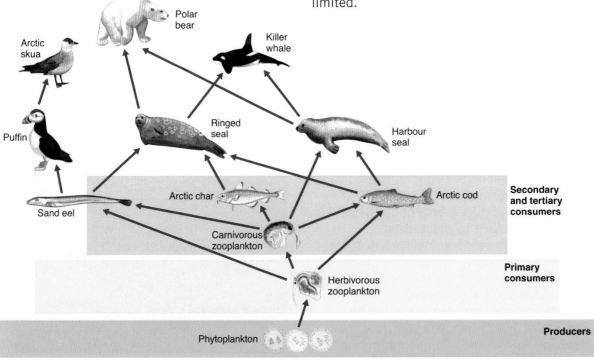

Figure 1 A food pyramid

Go Active

1 Build a model to show you understand how the nutrient cycle works. As it is a 'cycle', your model should be based on a circle. Explain your model to somebody else to make sure you understand it.

2 Show your understanding of a food web by firstly drawing a food chain for human beings. Then, develop this food chain into a food web. Label each animal (or human) as either a herbivore, carnivore or omnivore. Mark also the term **photosynthesis** and make sure you show how the **transfer of energy** alters throughout the food web.

How is the global distribution of large scale ecosystems influenced by climate?

There are several major biomes:

- Tropical rainforests
- The Savannah grassland
- Deserts
- The Mediterranean
- Temperate grasslands
- Deciduous woodland
- Coniferous forests
- The Tundra
- Mountains

> **Biomes**: large scale ecosystems across the world where the climate, vegetation and soils are all broadly the same within the area.

The small number of major biomes suggests that there are key factors which produce climax communities. Climate is the most important factor in determining the distribution of these large scale biomes. Relief, geology and soils are other important environmental factors.

Case Study – The tropical rainforest

- Tropical rainforests are found in equatorial countries such as Brazil, Congo and Thailand. They contain the most diverse range and highest volume of plant and animal life anywhere on earth.
- Tropical rainforests have hot and humid climates. It rains almost every day, totals often exceed 2000 mm a year. Temperatures are high with an average of 26 °C and there is a continuous growing season.
- The ecosystem has high inputs of energy from the Sun. Dead organic material and the hot, damp conditions on the forest floor allow rapid decomposition. This, together with rainfall, provides plenty of nutrients that are easily absorbed by plant roots.
- Nutrients are in very high demand from the rainforest's numerous fast-growing plants, so they do not remain in the soil long, staying close to the surface of the soil.
- Therefore, if the rainforest is cleared for agriculture, it will not make very good farmland, as the soil is not rich in nutrients. Also, once the trees (with their roots) are removed, the soil can be easily washed away.
- Vegetation is divided into layers – the **emergents**, **canopy**, **shrubs** and the **ground layer**. Trees are deciduous, growing and dropping leaves throughout the year. Species of tree include mahogany and teak.

Figure 2 Tropical rainforest

Go Active

1 Using the figures below, draw a climate graph like the one on page 69 to show the typical climate of the rainforest.

2 Annotate this climate graph and explain why it gives the most productive ecosystem on earth. Give at least two reasons.

3 Use the 'double bubble' diagram below to compare the rainforest ecosystem with that of the deciduous forest found in the British Isles. Compare their location, climate and examples of plant/animal adaptation.

You can repeat this with any of the other biomes.

	J	F	M	A	M	J	J	A	S	O	N	D
Precipitation (mm)	240	220	242	215	170	100	70	40	50	100	150	215
Temp (c)	27	27	27	27	27	27	28	28	28	27	27	27

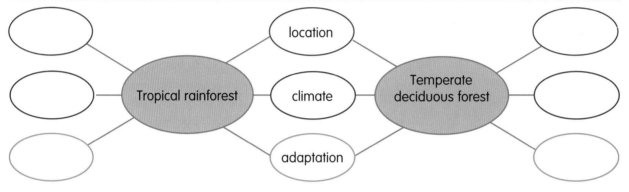

How are ecosystems managed?

In what ways do people use ecosystems?

The Essentials

How humans have affected ecosystems

Human activity has destroyed or changed parts of all ecosystems. Deforestation, agriculture, hunting and the introduction of new species have altered the food webs and nutrient and energy cycles over the entire earth.

- In the UK, much of our temperate deciduous forest has been removed for farming and to provide room for towns and cities. Today many forest areas have become tourist attractions.
- In the United States, much of the temperate grassland has been removed to grow cereals.
- In the Kalahari desert in Botswana, bushmen traditionally live by hunting and gathering the produce of the desert.
- In the Mediterranean, tourism and a demand for water puts the natural ecosystem under pressure.

How can ecosystems be managed sustainably?

- The fragile Tundra ecosystem is under pressure from oil exploration in Alaska.
- The Mediterranean ecosystem is under pressure from tourism, e.g. in Majorca, water shortages have led to tankers shipping water in from the mainland.
- The desert ecosystem in the USA is under pressure from intensive farming in California and people's excessive demand for water, e.g. for swimming pools.
- Coniferous forests are being felled in western Siberia to exploit the oil and gas reserves. Oil spills are polluting the land making regeneration impossible.

Sustainable: use of ecosystems aims to meet the needs of current generations, allowing the country to develop and improve the quality of life of its population, without compromising the needs of future generations.

Case Study – the tropical rainforest and its people

- Tropical rainforests are particularly under threat from human activity. Over the last 100 years, 50 per cent of tropical rainforest has been destroyed. Some scientists predict that most of the world's tropical rainforests will disappear in the next ten years.

Most rainforest areas are in poor countries which need the riches of the forest to help them develop. Rainforests are being cleared for many reasons:

- Farming – e.g. commercial cattle ranching.
- Mining – in the Amazon there are huge reserves of minerals such as gold, iron ore and copper.
- Roads – these improve access into the rainforest which makes it more easily exploited.
- Population pressure – land is needed in growing population for homes and new industries.
- Logging – expensive woods such as mahogany are exported across the world.
- Electricity supplies – hydroelectric schemes require the construction of huge dams.

Impact of human intervention

- **Employment** – jobs are created by logging, farming, mining and tourism.
- **Resettlement** – provides a better way of life for people who lived in shanty towns.
- Roads divide up parts of the rainforest and can cut off connections between different biotic and abiotic systems, e.g. a road can stop monkeys such as the Golden Lion Tamarin from travelling to gather food and, in turn, distribute seeds to re-sow plants in the forest.
- **Land clearance** – hardwood trees take many years to grow, so they can be difficult to replace. Once the nutrient cycle is destroyed it is often impossible for the natural ecosystem to develop again.
- **Fertile soils** – are quickly washed away when the forest is cleared for farming, mining or transportation. If soils end up in rivers, this can lead to flooding.
- **Loss of animal habitat** – trees are cut down and animals in them have to find somewhere else to live. Hence, deforestation can result in endangering animals and plant life, even causing them to become extinct.
- **Profits** – large scale farming, selling timber and minerals provide short term gains. Profits often go back to MEDCs or large companies who finance the logging and mining without benefiting rainforest communities.
- **Culture** – local people, who have lived in the forest for centuries, lose their homes and way of life.
- **Climate change** – burning of the forest adds carbon dioxide to the atmosphere and the removal of trees alters the carbon cycle in these areas.
- **Medicine** – many modern medicines have their origin in the rainforest and the removal of this ecosystem will prevent future discoveries.

Go Active

Complete a case study card on the rainforest:

1 Name of case study
2 Location of case study (including a map)
3 At least three reasons for the destruction of the rainforest
4 At least five ways in which humans have an impact upon the rainforest.

Case Study – Sustainable use of the rainforest

- MEDCs suggest that LEDCs should stop deforestation but it is the MEDCs that demand most of the timber. MEDCs have already deforested the temperate forests and exhausted some of their mineral resources. LEDCs need the profits from using rainforest resources to improve the lives of their populations.
- A compromise suggested by groups such as Friends of the Earth and the World Wide Fund for Nature, is to use the rainforests and all ecosystems sustainably.
- Indigenous rainforest tribes have used the rainforest resources in a sustainable way for centuries in a system of agriculture known as shifting cultivation.

Shifting cultivation

Shifting cultivation occurs in areas of the Amazon rainforest, central and western Africa and Indonesia. The method can be described as follows:

- A small area of land is cleared and the vegetation burned, the ash providing a source of nutrients.
- For a few years the soil remains sufficiently fertile for the tribe to grow crops and feed themselves.
- When productivity levels fall, the tribe will move on and clear another small area of forest.
- The original area will then regenerate, being small enough to receive nutrients and seeds from surrounding vegetation.
- As no lasting damage occurs, this method of agriculture is sustainable.

Along with other aspects of local rainforest culture and the traditional way of life, shifting cultivation is under threat from large scale clearance of the forests.

Sustainable management of the forest

Possible sustainable strategies include:

- **Agro-forestry**: this involves growing trees and crops at the same time. Farmers can then take advantage of the shelter provided by tree canopies, which prevents soil erosion, meaning their crops will benefit from the nutrients created from dead organic matter.
- **Selective logging**: trees are measured and only felled when they reach a particular height. This gives young trees a guaranteed life span and the area will regain full maturity after around 30–50 years. The Forest Stewardship Council links timber producers to customers and gives assurances that all timber sold comes from areas of sustainable logging.
- **Education**: this ensures that those involved in exploitation and management, including the general public in MEDCs, understand the principles behind their actions.
- **Afforestation**: this is the opposite of deforestation. If trees are cut down, they are replaced, thereby maintaining the forest canopy. Some mining companies must agree to do this before they begin exploiting rainforest resources.
- **Forest reserves**: areas of forest are protected from exploitation and maintained as natural environments. Some areas have a designated core where no human activity takes place, surrounded by a buffer zone where sustainable use of the forest is encouraged.
- **Monitoring**: this describes the use of satellite technology and photography to check that activities in the rainforest are legal and follow guidelines to ensure sustainability.
- **Ecotourism**: this, a rapidly growing form of tourism, is often encouraged and is highly profitable.

Go Active

Add a second card to the earlier case study card you created on the rainforest. This 'part two' card should have:

1 a definition of the term 'sustainable'
2 at least four ways in which the rainforest can be sustainably managed.

What are the likely consequences if ecosystems continue to be damaged?

What is the evidence that ecosystems are being used unsustainably?

The Millennium Ecosystem Assessment is a major report which has assessed the consequences of ecosystem change for human well being. More than 1,360 experts worldwide were consulted and their findings provide an appraisal of the condition of the world's ecosystems, as well as the basis for action to conserve and use them sustainably.

The assessment had four main findings:
- Over the past 50 years humans have changed ecosystems more rapidly and extensively than in any comparable period of time in known history. This has resulted in a substantial and largely irreversible loss in the diversity of life on Earth.

- The changes that have been made to ecosystems have contributed to substantial gains in human well being. However, this has also caused the substantial degradation of many ecosystems, which will significantly diminish the benefits offered to future generations.
- The degradation of ecosystems could grow significantly worse during the next half century.
- The challenge of reversing the degradation of ecosystems can be partially met but will require significant changes in policies and practices that are not currently underway.

Go Active

Practise being a newsreader about to read out the headlines of the day. Write a 30 second snippet to summarise the Millennium Ecosystem Assessment. You will need to think carefully about what it is and what it does – 30 seconds is not a long time!

What are the local and global consequences of unsustainable ecosystem use on people and the environment?

Human activities are damaging the biosphere – the balance of living and non-living things on planet Earth – with possible serious long term consequences e.g. climate change, desertification and increased flooding. Over-fishing in the North Sea, the destruction of fragile sand dunes ecosystems for caravan parks, logging in the rainforest are some examples of damage.

Scientists argue that ecosystems provide people with key services which include:
- Maintaining clean water in rivers
- Preventing soil erosion
- Reducing the risk of flooding
- Providing natural materials such as timber
- Producing foodstuffs such as honey and nuts.

Conservationists argue that maintaining key services is far more valuable for the long term well being of people on the planet than short term gains from unsustainable use of ecosystems.

Go Active

1 Draw spider diagrams for any two ecosystems to describe the immediate consequences of human actions on each ecosystem. Use one colour per spider diagram.
2 Then, using a different coloured pen, add the medium term consequences, i.e. what might happen over the next five years as a result of these immediate actions.
3 Finally, repeat the exercise for long term consequences, i.e. what might happen over the next twenty years.

Inside Information

Information which is often presented in the form of maps, diagrams, graphs, newspaper articles or cartoons is an important part of Geography. You therefore need to develop your skills of interpretation so that you can maximise your marks in the examination.

Before you read the question you should study carefully all information – you may think it is useful to scribble a few notes next to the diagram. In your answer, use evidence from the diagrams/graphs and give examples/figures. Never simply make a list, unless this is demanded in the question.

a) Study the climate graph in Figure 3.

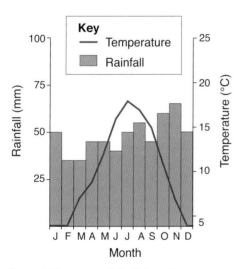

Figure 3 Climate graph for Salcey Forest

i) What is an ecosystem? [2]

Student answer

A system that joins✓ animal and plant life together.

Examiner's comments

An ecosystem is a natural system in which plants, animals and non-living things interact with each other. I would give this candidate 1 mark only for the idea that animals and plants are linked.

ii) Describe the changes in average temperature throughout the year in Salcey Forest. Explain how these changes affect the vegetation of the forest. [4]

Student answers

Student A

The temperature rises through the summer months and then falls in the winter.✓ This means that vegetation is low in winter and high in summer.✓

Student B

In the winter months it's a very low temperature and as it goes into the summer months, the temperature increases✓ to 18 °C.✓ This will affect the growth of vegetation because the temperatures rise above the growing season temperature of 6 °C✓, so the plants will begin to grow. ✓

Examiner's comments

There is simply not enough detail in student A's answer to score four marks. In any point marking scheme you must look to make enough points to attain all the marks available. Student B's answer is weak but s/he uses the information and gives figures from the graph, which just makes four points.

iii) Describe how human use of Salcey Forest is likely to affect the forest's ecosystem. [3]

Student answer

Humans could cut down some of the trees which would get rid of animal habitations.✓ Also, they could dump their rubbish in the ponds which could harm/poison the fish.✓

Examiner's comments

This is another very weak answer but the student makes two separate points and, although very general, I would give this 2 marks. However, rather than looking to make three points, the candidate merely filled the space available for the answer.

EXAM SPOTLIGHT

Study the information below:

THE AMAZONIA LODGE, BRAZIL

The Lodge is reached by boat along the River Negro.

Native guides provide education about local plants.

All sewage is recycled.

There is an observation tower and a canopy walkway providing views of the area.

Trees cut down to build the hotel and walkway have been replanted.

Native people are employed at the Lodge.

Using the information above and your own knowledge, explain how the development of ecotourism in Amazonia contributes to sustainable development. [5]

Why does the nature of tourism differ between one place and another?

What are the factors, physical and human, that affect the nature of tourism?

 **The Essentials**

Tourism: any activity where a person voluntarily visits a place away from home and stays there for at least one night.	**Domestic tourism**: where people visit places in their own country.	**International tourism**: when people visit other countries.

We usually think of tourism as going on holiday, but there are many other reasons why a person might visit a place. These include:
• to attend a sporting event
• to visit friends or celebrate an event like a wedding
• to improve health, e.g. visiting a health spa
• for business, e.g. attending a conference
• for prestige – to be able to tell your friends about it.

Tourism is a service industry. It provides pleasure for some and a source of income for others through a wide range of jobs. Some people are employed directly, such as a chef in a restaurant or an airline pilot, while many others are employed indirectly, like a police officer or an insurance agent.

Tourist destinations develop because they are attractive to people. Here are some examples:
• Cities contain shops, restaurants, bars, theatres, museums, famous buildings and night life. Top tourist destinations include London, Venice and Las Vegas.

• Coastal areas include beaches for sunbathing, sea for water sports and an attractive environment for activities like walking. Famous coastal resorts include Blackpool, Marbella and Santa Barbara in California.
• Mountains provide spectacular views, rock climbing, walking, white water rafting in fast flowing mountain rivers and skiing. Famous destinations include the Pennines, the Alps and the Himalayas.
• Climate is an important factor. The Mediterranean has hot, dry summers which are perfect for beach holidays, while snow in the Alps provides ideal conditions for skiing.
• Resorts such as Center Parcs and Disneyland have attracted millions of visitors in recent years.

Tourism is the world's largest and fastest growing industry, with a turnover of around US $700 billion. Until recently it was an activity which was largely concentrated in the richer, MEDC countries of the world but it is increasingly becoming an important source of income in LEDCs.

 ### Case Study – Pembrokeshire Coast National Park
The Pembrokeshire Coast National Park boasts some of the most spectacular scenery and diverse wildlife in Britain. National parks conserve the natural features and cultural heritage of an area, whilst giving people access so that they can enjoy the natural environment.

Attractions
These include:
• spectacular scenery in the UK's only coastal National Park
• 186 miles of the Pembrokeshire coastal path
• heritage sites including 51 castles
• access to resorts such as Tenby and Saundersfoot
• access to the Preseli mountains
• over 50 beaches, many of them 'Blue flag' beaches
• access to Oakwood theme park
• exposed to the southwesterly prevailing winds, which provide ideal conditions for surfers.

Go Active

1 Group the physical and human attractions of Pembrokeshire in the Venn diagram. Are there any physical attractions which have been enhanced by people? These should go in the middle of the diagram, e.g. the 'Blue flag' award for beaches.

2 Use the outcomes shown on your Venn diagram to help you decide whether you think physical or human attractions are most important to Pembrokeshire.

3 Carry out your own research. Find four reasons why Disneyland (or another popular tourist destination) has become one of the top international tourist destinations.

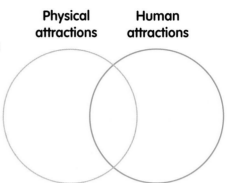

In what ways and why is tourism changing?

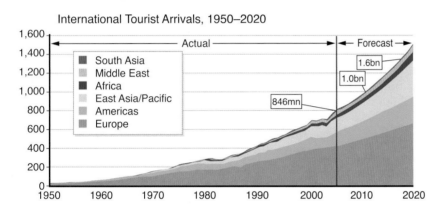

Figure 1 International tourist arrivals

Go Active

1 Re-draw the table to put the countries in order of largest to smallest amounts generated by tourism.

Country	US $billion (2007)
Spain	58
USA	97
UK	38
Austria	19
Germany	36
France	54
China	42
Turkey	18
Italy	43
Australia	22

The Essentials

Many factors have contributed to the growth of the tourist industry since the 1950s, including:

- an increase in the amount of paid holidays and higher salaries
- television travel programmes have raised people's expectations
- increased life expectancy and early retirement mean there are a greater number of older people travelling
- travel has become easier, with air fares even to distant destinations becoming more affordable and car ownership increasing
- the growth of the package holiday industry has made booking easier and holidays more affordable
- the growth of the internet and modern telecommunications has allowed people to easily make their own travel and accommodation arrangements.

There have also been big changes in the distance people are prepared to travel, the time of year people take holidays and the nature of holidays taken.

- Distant places like Florida, Kenya, Thailand and even Antarctica have become potential tourist destinations.
- Ski resorts, once exclusive to the rich, attract increasing numbers of tourists.
- There has been massive growth in 'short breaks', e.g. city breaks.
- There has been a growth in the number of purpose built resorts, such as Center Parcs, which houses extensive indoor facilities that are not dependent on weather.
- There has also been a growth in business tourism, including international business meetings and weekend motivational conferences.

Life cycle model

Geographers have suggested a model for the life cycle of a tourist destination. It is divided as follows:

- **Stage 1: Discovery** – few tourist facilities, visited by a small number of adventurous travellers, e.g. Guatemala.
- **Stage 2: Development** – tourist facilities and accommodation is built, numbers of visitors increase, e.g. Thailand.
- **Stage 3: Consolidation** – tourist facilities and hotels support large numbers of tourists, e.g. Cancun, Mexican Riviera.
- **Stage 4: Stagnation** – tourist numbers peak, large numbers of visitors cause environmental damage and the resort becomes less fashionable, e.g. Spanish Costas.
- **Stage 5: Decline** – numbers decline as tourists seek new, unspoilt and exciting destinations, e.g. Barry, Rhyl and Morecambe

OR

- **Stage 6: Rejuvenation** – investment, advertising, new facilities and attractions draw in new visitors, e.g. Majorca.

Go Active

Some geographers have suggested a seventh stage to this model, called **Sustainability**. Currently this stage has not been implemented, but if you were in charge of tourism in a region, describe and explain three ways in which successful tourism could be sustained once rejuvenation has taken place. If you read further into this chapter, you will encounter several examples of and references to sustainable tourism which may help you.

Case Study – The traditional British seaside resort

The growth of tourism in the UK was largely based around coastal resorts.

- 1700s – a period of discovery, with the growth of spas and early seaside resorts.
- 1800s – mass tourism developed with the creation of railway connections from new industrial cities to rapidly growing seaside resorts such as Blackpool.
- 1950s – saw consolidation and a peak in the number of visitors.
- 1970s – a period of stagnation, due to competition from new Mediterranean resorts such as Benidorm and the growth of package holidays.
- 1990s – decline was seen due to the British weather and often shabby tourist attractions, e.g. Margate.
- 1990s – rejuvenation was also seen with resorts such as Blackpool, with its illuminations and reputation as a destination for hen and stag parties. Other resorts such as Newquay in Cornwall have diversified and now attract world famous surfing championships.

What are the impacts of tourism?

What are the impacts of the development of tourism on people, the economy and the environment in one MEDC region and one LEDC region?

The Essentials

Tourism Benefits	Tourism Costs
Provides jobs	Jobs are often low paid and temporary
Brings foreign exchange	Local culture is destroyed
Provides wealth that can be invested in services such as health and education	Fragile ecosystems, e.g. sand dunes, are destroyed
Preserves local culture	New roads, airports and increased traffic cause pollution and environmental damage
Environments can be protected	
New facilities are built	

Case Study – Tourism in a MEDC – the Costa del Sol

The Costa del Sol is located in the south of Spain. Until the 1960s the main jobs in the region were farming and fishing and living standards were low.

Since then the region has been transformed into one of Europe's most popular tourist destinations with over 7 million visitors per year.

Attractions

Physical	Human
Dry, warm climate	Historic towns
Long, sandy beaches	Spanish culture
Warm sea	Well developed roads and airport
Rugged mountains	Entertainment and nightlife

Impact

Benefits	Costs
• There is a modern airport at Malaga.	• Many jobs are low paid, temporary and low status.
• New roads have been created, e.g. N340.	• Unemployment in the region has recently risen to 30 per cent.
• 70 per cent of people work in tourism.	• There is increased crime, vandalism, drunkenness and violence.
• Living standards have improved.	• Local culture, music and dancing have become merely tourist shows.
• Traditional crafts, e.g. lace making, have found a new market.	• Many high rise hotels, built in the 1960s, look shabby.
• Some beaches have blue flag status.	• Traffic congestion is acute.
• Nature reserves have been set up.	• The sea has become polluted from litter and sewage.
	• Water supplies are under pressure.

Sustainable tourism?

By the late 1990s the resorts of the Costa del Sol were in decline. Prices were high and the media image was poor. Since then, Spanish authorities have been partially successful in halting the area's decline and rejuvenating tourism in the region.

• Further high rise building has been banned and new building must be in line with the traditional Spanish courtyard style.

• Resort centres have been pedestrianised and planted with trees to improve their image.
• Bypasses to the resorts have been built to ease traffic congestion.
• The building of golf courses and luxury villas has been encouraged. This has attracted different sorts of holiday maker in an attempt to improve the region's image.
• Cost cutting winter breaks have been offered to the retired and elderly.

Case Study – Tourism in a LEDC – Kenya

- Tourism has become a global industry and LEDCs are keen to attract tourists in order to promote development.
- Kenya was one of the first LEDCs to develop its tourist industry.
- As a former British colony, it is English speaking, which certainly helped its appeal to tourists. Kenya earns around US $500 million per year more from tourism than for its export of tea and coffee put together.

Attractions

Physical	Human
Hot climate	English speaking
Long, sandy beaches	Different cultures
Warm sea and coral reefs	Well developed roads and airport
Safari parks	Cheap cost of living

Impact

Benefits	Costs
• Half a million people are employed.	• Jobs are low paid and temporary.
• Living standards have improved, with more schools and hospitals.	• Foreign multinational companies own 80 per cent of the hotels and travel companies. Most of the profits go back to MEDCs.
• Kenya's infrastructure has improved.	• Nomadic tribes are forced to settle. Traditional ways are lost and they dance for tourists to make a living.
• Traditional culture and skills have been retained.	
• Local tribes are able to make money, e.g. the Maasai tribe sell handicrafts.	• Alcohol and standards of dress offend Kenyan Muslims.
• Safari parks protect animals from poachers and prevent extinction.	• Sex tourism is common on the coast.
	• Coral reefs are damaged.
	• Local fishermen lose their livelihood, as waters are overfished.
	• Game park buses cause soil erosion and alter animal behaviour.
	• Savannah hotels use up precious water.

Sustainable tourism?

Although tourism is a relatively new industry in Kenya it has already reached the point of decline. Political instability, violent crime, harassment of tourists, over commercialism of safari parks and environmental degradation has turned away some tourists. The Kenyan government is now acting to try and halt this decline by:

- limiting the use of existing marine and game parks
- promoting sustainable tourism where power supplies are limited and visitors are in small groups
- using local people as guides
- limiting the height of new hotels to that of the trees
- asking tourists to observe local customs.

Go Active

1 Create a case study card for the Costa del Sol and Kenya. Whereas your case studies normally need to fit onto one card, this one can expand onto two because it will compare and contrast the two locations.

2 Choose two coloured pens, one for the Costa del Sol and one for Kenya. Construct your case study card to cover the following areas:

- a location map for each place
- a brief introduction to each place
- five attractions for each destination
- five impacts of tourism on each place
- two ways in which each place is attempting to manage tourism sustainably.

3 Write the information for both places next to one another, in different coloured pens, so that you can easily compare and contrast.

How can tourism be developed in a sustainable fashion?

The Essentials

People responsible for tourist sites are looking to manage them in sustainable ways. Sustainable methods include limiting visitor numbers to prevent damage, as used at Stonehenge and the Pyramids of Egypt.

Management strategies in national parks, such as the Lake District, include:

- providing park and ride schemes to encourage people to leave their cars outside the park
- reinforcing footpaths and encouraging people to use marked paths
- developing honeypot sites where tourists are concentrated and problems such as litter can be more easily managed
- demanding that areas of quarrying are landscaped after use.

Honeypot sites

This strategy involves identifying popular locations and allowing development. Bigger car parks are built, public toilets are provided, hotels and cafes allowed and access roads improved. The damaging environmental effects of tourism can be more easily managed when tourists are concentrated in a small area. Thus, some areas are sacrificed to protect others.

Ecotourism

Ecotourism or green tourism aims to give local people jobs, for example as guides, whilst protecting the environment. It improves people's quality of life without destroying their culture and way of life. Ecotourists travel in small groups and often visit reserves where the scenery and wildlife is protected and managed.

Honeypot site: a place which has attractive scenery or historic interest which attracts tourists in large numbers, 'like bees to a honeypot'.

Ecotourism: a sustainable form of tourism. It tries not to damage the environment and respects local culture and customs.

Case Study – Ecotourism in the rainforest

Ecotourism is rapidly becoming a leading method for developing countries to bring in foreign money.

- In this ecolodge, groups of no more than twenty people stay in lodges made of local material.
- Guests are not allowed to take any foodstuffs in from outside the forest, to prevent litter and contamination of the ecosystem.
- Compost toilets recycle human waste.
- Guests are taken on rainforest walks where they are educated about the wildlife.
- Local tribe people sell locally made handicrafts to make a small amount of spending money.
- Money spent directly in the local economy helps give economic value to rainforest preservation.
- The locals, along with the government, can see the importance of keeping the forest intact.

Figure 3 An ecolodge in the Peruvian rainforest

Go Active

1 Create a case study card for ecotourism in the rainforest.
2 Check your recall and understanding of your tourism case study cards by trying to explain their content to one of your classmates. Try doing this once using the card itself and then by giving the card to your classmate to check you have remembered everything.

Aiming for A*

When you read a question, it is important to think about what you are being asked. If it's worth eight marks, remember to think about the structure of your answer and plan one that will address all parts of the question. You will achieve an A* if you include detail, use correct geographical terminology and case studies. You must know your stuff but also be able to use your knowledge where appropriate.

Sample question

For a tourist region you have studied describe how successful different attempts have been to develop sustainable tourism. [8]

Student answer

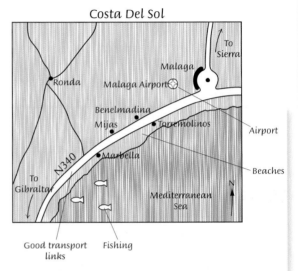

Costa Del Sol

The Costa del Sol is located in the south of Spain. It has been a popular holiday destination for many decades and was one of the first regions in the Mediterranean to be developed. People were attracted by the hot, sunny climate, sandy beaches and warm sea. In the 1990s the reputation of the region suffered and tourism began to stagnate and even decline in some resorts. Tourists began to go to new, trendier places. The Spanish government have made attempts to rejuvenate tourism. Old hotels have been demolished and newer, more upmarket hotels built. Marbella is marketed as a designer resort. The government has encouraged the Spanish culture and shops, restaurants and hotels reflect this. In an attempt to clean up beaches the government are trying to achieve the EU Blue flag award. They have also put a lot of effort into cleaning up litter and machines clean beaches every morning. VAT was reduced to six per cent to reduce the price of holidays and new attractions such as golf courses built.

Examiner's comments

This is a good answer. It is well structured, contains detail, uses correct geographical terminology and a relevant case study. However, the candidate has not fully answered the question since s/he has not commented on the success of the strategies, hence this answer is worth 7 marks.

Mark Scheme

Level	Description
1 (1–3 marks)	Limited knowledge of named region. Little understanding of the concept of sustainability, although candidate should identify some measures to promote tourism.
2 (4–6 marks)	Clear knowledge of named region. Clear understanding of the concept of sustainability and answer describes a range of measures adopted to promote sustainable tourism.
3 (7–8 marks)	Clear and detailed knowledge of named region. Clear understanding of the concept of sustainability. Describes a range of measures adopted to promote sustainability and begins to evaluate the success of these strategies.

EXAM SPOTLIGHT

a) For a region you have studied outline how both physical and human factors have encouraged the development of tourism. [4]

b) Imagine you are the leader of a government in an LEDC. Many people are critical of the development of tourism in your country. Write a speech to the voters in which you outline your views on the issue. In your speech you should summarise the possible benefits and costs of tourism and state whether or not you are in favour of its continued growth. [8]

In what ways are European city centres being renewed?

How are European city centres changing?

The Essentials
Central business district
The CBD is characterised by:

- high multi-storey buildings
- expensive land values
- department stores and specialist shops like jewellers and fashion stores
- modern shopping malls
- religious and historical buildings, museums and castles
- offices, banks and government buildings
- bus and railway stations, multi-storey car parks.

The CBD is located where roads and railways converge and is the most accessible part of the city.

The **central business district (CBD)** is the commercial and business centre of a town or city where the land values are highest.

Figure 1 The central business district of Cardiff

Change in the central business district

Some CBDs have become run down as shops and businesses face increasing competition from out of town shopping centres, as in Dudley CBD with the construction of Merryhill.

Measures taken by city authorities to revitalise CBDs include:

- pedestrianisation – making the area safer and more pleasant for shoppers
- improving access with better public transport and improved car parking

- converting derelict warehousing into trendy shopping units, restaurants and museums
- improving public areas like parks, statues, street furniture and more greenery to make them more attractive
- building undercover shopping malls such as St David's in Cardiff and Cabot's Circus in Bristol.

Today, city centres are increasingly seen as places of entertainment and leisure with cinemas, theatres, restaurants and bars.

Go Active

1 Write in the style of a Facebook status update to explain why Cardiff is an attractive environment for shopping. For example, Stacey McCabe thinks Cardiff is great for shopping because …

2 Visit your local town or city and take photographs of the features listed.

Remember, local places you know well are good for using as case studies.

- pedestrianisation
- improved public transport
- new car parking areas
- new shops, restaurants
- statues
- street furniture
- greenery

Case Study – Manchester's CBD

- The IRA bombing of Manchester city centre in 1996 has greatly influenced the regeneration efforts of the city in subsequent years.
- The aim was to provide an environment which would be attractive for all users of the city centre, themed around safety and reassurance for the users.
- The number, quality and range of shops were increased, including the improvement of the Arndale shopping centre. Hanging baskets and shrubs improved the appearance of the centre.

- Retailers worked with the police to reduce retail crime. A large CCTV system was installed with over 400 cameras across the city centre. A rapid response team working against vandalism has been set up and street crime wardens appointed.
- Seats were removed from parts of the centre and beggars and undesirable people associated with petty crime and nuisance were given less opportunity to loiter in the area.
- Increases in housing sought to repopulate the city centre.

Go Active

1 Use two different coloured pens to highlight the Manchester case study: one colour for reasons why regeneration was needed and the second colour for the ways in which the city has been regenerated.

2 Annotate the photograph of Manchester CBD in Figure 2, giving examples of ways in which the centre of the city has been made more attractive. Use the information in the text and your own observations.

Figure 2 Manchester's CBD

The inner city

The inner city is also known as the **twilight zone**. It is located next to the CBD and consists of older terraced houses and newer tower blocks.

Changes in the inner city

Many inner city areas have suffered from environmental, social and economic problems. This culminated in unrest and riots during the 1980s, e.g. Brixton in London. The government response was to redevelop and regenerate inner city areas.

> **Brownfield site**: a site which has previously been built on but has become derelict needs redevelopment.

Redevelopment involves improving the physical environment by clearing old buildings and developing new businesses and housing on the brownfield sites. This helps create jobs, raise the quality of life in the inner city and attract people back to the area; a process known as **regeneration**. Some of the most ambitious regeneration schemes have been based in former dockland areas, e.g. Cardiff Bay and the London Docklands.

Case Study – Cardiff Bay: Europe's largest waterfront development

Cardiff Docks was the world's largest coal exporting port in the early 1900s. The decline of the coal mining industry in south Wales led to its dereliction and a spiral of decline in nearby communities like Butetown.

The Cardiff Bay Development Corporation (CBDC) was set up in April 1987 to regenerate the derelict dockland areas. It was part of the government's Urban Development Programme which aimed to regenerate deprived and run down areas of British inner cities. The mission statement of CBDC was as follows:

'To put Cardiff on the international map as a superlative maritime city which will stand comparison with any such city in the world, thereby enhancing the image and economic well being of Cardiff and Wales as a whole.'

The five main aims were as follows:

1 To provide a superb environment in which people will want to live, work and play.
2 To reunite the city of Cardiff with its waterfront.
3 To create job opportunities and reflect the hopes and aspirations of the communities in the area.
4 To achieve the highest standard of design and quality in all types of development.
5 To establish the area as a recognised centre of excellence and innovation in the field of urban regeneration.

Today, Cardiff Bay is home to a number of leisure attractions such as Techniquest, the Welsh Assembly and the Wales Millennium Centre. The construction of the barrage is one of the largest engineering projects in Europe. Completed in 1999, it has created a 500 acre freshwater lake with eight miles of waterfront and, it is hoped, will stimulate the future development of the bay as a tourist and leisure destination.

Go Active

1 Create a mind map to summarise the three key areas of information on Cardiff Bay. Use one colour/branch for why redevelopment and investment was needed, a second colour/branch for the five aims of redevelopment and a final colour/branch for Cardiff Bay today.

What are the effects of city centre changes on their day and night time geographies?

The Essentials

Traffic

Traffic in urban areas has increased dramatically. Increased affluence has led to increased car ownership and use. The movement of people away from central areas of cities has led to increased commuting.

> **Commuter**: a person who lives in a small town near a city and travels to the city for work.

Most city centres were not designed for the heavy use of cars. This has led to accidents, congestion and pollution. These problems are particularly acute during the morning and evening rush hours. Time wasted in traffic jams is expensive for businesses and exhaust fumes add to global warming.

In recent years, employers, the government and local councils have tried to reduce the problems of traffic congestion by:

- introducing **flexi-time** to spread the traffic load
- improving **public transport**, e.g. supertrams in Sheffield
- implementing traffic management schemes including the creation of **bus lanes**, developing **park and ride** schemes and encouraging people to **cycle** by creating cycle lanes
- introducing a **congestion** charge, as in London, for vehicles entering the centre on weekdays.

Crime and antisocial behaviour

Cities have increasingly become leisure and entertainment centres. In the day people are attracted by the shops and tourist attractions, at night by theatres, pubs and clubs.

> **Antisocial behaviour**: any aggressive, intimidating or destructive activity that damages or destroys another person's quality of life.

The high concentration of people in the centre has led to increases in crime and antisocial behaviour. In the day, there has been an increase in petty crimes such as shoplifting and pickpocketing. At night, the large numbers of young people drinking alcohol has led to noise, violence and antisocial behaviour.

Increasingly city authorities have looked to adopt measures to tackle the problem, such as **Designated Public Places Orders (DPPOs)** which give police the power to stop the drinking of alcohol in designated areas and seize alcohol from nuisance drinkers in a bid to reduce disorderly behaviour caused by drunkenness.

The changing nature of inner-city society

A recent trend in many city centres has been an increase in the residential population of the CBD and in the attractiveness of many formerly run down inner city areas. This is partly to avoid the travel costs and delays of accessing the centre and partly due to government regeneration policies. Another important factor is the growing proportion of single people of all ages in society. The city centre, with its shopping and entertainment functions, is an attractive environment to inhabit for many.

The movement of more affluent people into the inner city and the improvement in housing quality is termed **gentrification**. The process has significant effects on central city areas:

- The average income of communities increases whilst the number of children decreases.
- Former lower income residents, due to increased rents and house prices, may be forced out of the area.
- Factories may close as land is redeveloped for housing, so jobs are lost.
- New businesses, such as restaurants catering for more affluent consumers, tend to emerge. This increases the city centre's appeal to more sophisticated migrants.
- Local residents may not be able to afford the new attractions.

Population movements have also changed the social and economic mix of people in cities. This has made many urban areas increasingly multicultural. Migrants from similar ethnic and cultural backgrounds often group together, for reasons including the following:

- People choose to live close to others with the same background, religion and language.
- People want to live close to services that are important to their culture, e.g. places of worship.
- People from the same ethnic background may also be restricted in where they live by house prices and prejudice from the host population.

The multicultural nature of modern cities makes them vibrant places to live and visit, although the segregation of people into different ethnic and cultural areas can sometimes lead to tension and even violence.

Strategies aimed at reducing the problems created by multicultural cities are aimed at equal access to services rather than forcing people to mix. Examples of this include providing leaflets in a variety of different languages and recruiting police officers to reflect the ethnic backgrounds they police.

What are the current patterns of retailing in European cities?

Where does retailing occur in the city?

The Essentials

Shopping hierarchy

Shops can be placed into a hierarchy based on the services they provide.

Retailing: the sale of goods, usually in shops, to the general public for their personal use.

- At the bottom of the hierarchy are small shops selling low order, convenience goods such as bread and milk. These are located in suburban high streets and local centres.
- At the top are shops selling high order goods such as furniture and electrical goods. These are located in the central business district.

Since the start of the 1980s there has been a movement of retailing to locations at the urban–rural fringe. Regional shopping centres and out of town locations have replaced the CBD in today's modern shopping hierarchy, e.g. Meadowhall and Merry Hill.

Shopping centres at the urban–rural fringe include:

- **superstores** such as Tesco, usually with a petrol filling station
- **retail parks** containing large stores selling goods such as furniture, garden supplies, carpets, DIY and clothing
- **regional shopping centres**, consisting of a large car park and undercover shopping mall which usually contains two or three major chain stores, e.g. John Lewis, and many smaller, national shops such as the Body Shop, restaurants and cinema screens
- **outlet villages** where chain stores sell discounted lines, often the previous season's fashions.

Case Study – Cribbs Causeway

Cribbs Causeway is a road just north of Bristol which has given its name to a large out of town shopping centre. It is accessible via the A4018 from Bristol and the M5. The Mall, with around 14 million visitors a year, is one of the two major shopping centres in the Bristol area, the other being Cabot's Circus.

- The Mall opened on 31 March 1998 and comprises of 135 shops on two levels.
- At its centre is a fountain, the money thrown into which is donated to local charities.
- Above the fountain is an extensive food court, which hosts many cafés and restaurants such as KFC and Bella Italia.

- The Mall has both lifts and staircases which take shoppers from floor to floor. It also contains the Venue, an entertainment complex featuring a VUE cinema and Hollywood Bowl.
- It is occupied by two **anchor stores**, John Lewis and Marks & Spencer, along with fashion retailers, jewellers and audio–visual stores.
- There are over 7000 free parking spaces and a bus station, from which services run around the Bristol area and as far afield to Weston Super Mare, Bath and Cwmbran.

The Cribbs Causeway complex consists of two further retail parks with shops such as Currys, Halfords, Toys 'R' Us, DFS and the Cribbs Business Centre.

Go Active

You are the general manger of Cribbs Causeway. Due largely to the redevelopment of Bristol city centre and a new shopping area called Cabot's Circus, you have to lead a publicity campaign to encourage more shoppers to visit your stores. Plan what features and elements you would highlight in your campaign. If you can, present your campaign to friends.

How is retail changing and what effects does this have on people and the environment?

The Essentials

Out of town shopping

Since 1980 it has been estimated that four fifths of all new shopping floor space has been created in out of town sites. The first large regional shopping centre to be developed in Britain was the Metro Centre in Gateshead.

Other land uses that find the urban–rural fringe an attractive location include housing, golf courses, allotments, business parks and airports. This mixture of land use often causes conflict, as different groups have different needs and interests.

Out of town shopping centres have developed for the following reasons:

- There are higher levels of car ownership.
- Their position near main roads makes the delivery of goods easier and gives convenient access to shoppers.
- There is plenty of open space for car parks, which attracts motorists. Unlike in city centres, there are no parking problems or traffic congestion.
- Land values are lower than those in the CBD. This allows shops to use large areas of floor space and keep the price of their goods down. Being so large, shops can stock a large volume and a wider range of goods.
- They are near suburban housing estates which provide a workforce, particularly as many employees are female, part time and have to work late most evenings.

83

The development of out of town shopping has caused problems for town centres including a loss of customers and the closure of small local stores. Vacant premises are often filled by discount and charity shops. Also out of town stores are less accessible to the elderly and those without cars. Environmental groups often oppose new developments because it means a loss of green land and an increase in pollution from cars. The closure of city centre stores and the concentration of commercial activity at the outskirts has led to 'hollow' centres in some cities. This effect was first described in the USA as the donut effect.

Internet

Online shopping has increased rapidly in recent years. Internet shopping is popular because it is convenient and often cheaper. Other advantages and disadvantages include the following:

Advantages	Disadvantages
Customers can buy products not available locally at cheaper prices.	Not everyone, particularly the elderly, have internet access.
Customers can buy from the comfort of their home, regardless of their mobility.	Goods may not be as expected when delivered and it may be difficult to return them.
It is less time consuming.	Shops lose trade and subsequently jobs.
Traffic congestion is reduced.	More delivery vans increase traffic congestion and pollution.
Jobs are provided for those delivering products.	

Case Study – Amazon.co.uk

Jobs boost as web warehouse opens
16 April 2008

A distribution centre for the online retailer Amazon covering an area the size of 10 football pitches has been officially opened in Swansea Bay.

The site in Jersey Marine is expected to create 1200 full-time jobs over five years, and 1500 seasonal jobs. First Minister Rhodri Morgan was at the opening and he called it a "powerful shot in the arm" for the Welsh economy. However, the Unite union has warned against Wales relying on service sector jobs. The warehouse is the company's fourth distribution centre in the UK and its largest.

"To have such a big name in e-commerce set up a major European base in Swansea Bay is an outstanding achievement for Wales", said Mr Morgan

BBC News website: http://news.bbc.co.uk/1/hi/wales/7349546.stm

Go Active

1 Do a quick survey of ten people (preferably from different age groups) and find out if they use the internet for shopping. Ask them what they buy, how often, why they use the internet and any other interesting pieces of information on internet shopping habits you can think of.

2 What do these results tell you? Why do you think these results have emerged?

3 Talk to someone who doesn't drive *or* doesn't have the internet at home. Find out from them how the development of out of town shopping centres *or* an increase in internet shopping has affected them. Are there positives and negatives for them?

4 Finally, complete the table below which demonstrates ways in which the environment is affected by these changes in retailing.

Positive effects of an increase in internet shopping	Negative effects of an increase in internet shopping	Positive effects of the development of out of town shopping centres	Negative effects of the development of out of town shopping centres

How do changes in European consumer choice have a global impact?

What are the impacts of increasing consumer choice on people in developing countries, and on the global environment?

Until the 1960s people generally ate a small range of seasonal food that had been grown in their home country, often in the local area. People now demand a range of foods which are grown all year round, regardless of growing season. Much of the produce on supermarket shelves is now imported. This has several positive aspects, including:

- a market for farmers in LEDCs
- employment for factory workers in LEDCs and MEDCs

- a greater range of products often at cheaper prices for European consumers.

Retailing is an excellent example of an industry that has been transformed through globalisation. Imported clothing, for example, has transformed the fashion industry, providing affordable clothing for consumers in MEDCs and employment for many thousands in LEDCs. However, many of these people work in sweatshop conditions.

Case Study – Wal-Mart

Wal-Mart began in 1962 when Sam Walton opened his first store in Arkansas, USA. More stores opened across the USA and more recently across the world, e.g. Mexico, Japan, Brazil and the UK where it is called Asda. Wal-Mart sells a variety of goods, food, clothing and electrical goods and has become the biggest retailer in the world, with 8000 stores employing over 2 million people.

Advantages of Wal-Mart	Disadvantages of Wal-Mart
Every new store creates around 500 jobs. Wal-Mart donates millions of dollars to improve schools, healthcare and the environment in the country in which it locates.	Workers in China work for only $1 per hour. Some companies that sell to Wal-Mart have long working hours and poor working conditions, e.g. Beximco in Bangladesh supply Wal-Mart with its clothing.

Case Study – Sweatshops

- Sweatshops are working environments in which conditions are considered to be difficult or dangerous and usually where the workers have few rights. It can include exposure to harmful materials, extreme temperatures, or abuse from employers.
- Sweatshop workers often work long hours for little pay, regardless of any local laws on overtime or wages. Workers are often women and there are many reports of child labour being used.

- They can exist in any country, not just LEDCs.
- Sweatshops usually employ low levels of technology but produce many different goods such as toys, shoes, clothing, and furniture.
- Defenders of sweatshops claim that people choose to work there because they offer substantially higher wages and better working conditions than their previous jobs of manual farm labour. They are an early step in the process of economic development.

The environmental cost of globalisation

As well as concerns over the exploitation of workers, many are concerned at the environmental cost of globalisation. Transporting goods produces CO_2 and the distance over which food is transported to the market is called food miles. The higher the food miles, the more CO_2 is produced and the greater the contribution to global warming. The amount of CO_2 produced during growing and transporting food is known as its carbon footprint.

Fair trade

The farmers in LEDCs who produce products such as coffee and cocoa usually make very little money. The greatest profits are made by the multinational companies who dominate the trade. Fair trade is a way of doing business with these farmers that ensures they get a fair wage for their work. It aims to provide minimum wages, safe working conditions, restrictions on child labour, protection for the environment and improved schools and healthcare. Many fair trade products are now on supermarket shelves and the sector has seen very rapid growth in recent years.

Go Active

This theme contains many keywords and definitions. To revise these, create a set of empty pieces of card (about as big as a sticky note). Onto half of them write the keywords (normally found in bold throughout this chapter) and onto the others write the definitions. Mix up these pieces of card and try to match up each keyword with its definition.

Understanding maps

Maps are an essential tool for geographers. It is almost certain that in your GCSE examination, at least one question will contain an Ordnance Survey map and it is likely that other questions will contain simple maps and sketch maps. It is therefore vital that you have basic map reading skills and can do the following:

- Read symbols, give grid references and direction, measure distance, use scales and understand contour lines.
- Describe geographical features shown on a map, e.g. physical features such as river valleys and human features such as communications.

Sample question

Study the sketch maps.

i) Calculate the approximate distance you would travel if you completed one circuit of the ring road. [1]
ii) Describe the distribution of newsagents in this city. [3]
iii) Compare the distribution of shoe shops and newsagents in the city. [3]
iv) Explain the distribution of both newsagents and shoe shops in this city. [4]

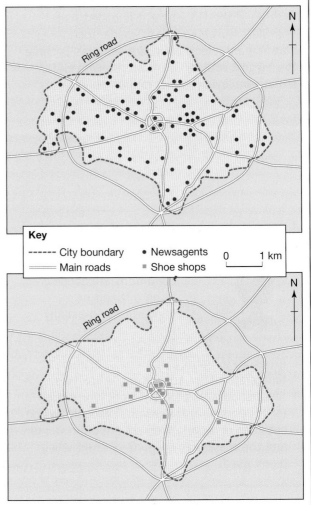

Key
------ City boundary • Newsagents
===== Main roads ■ Shoe shops
0   1 km

Student answer

i) 14 km

ii) *There are a lot of newsagents along main roads.✓ They are mainly evenly✓ distributed through the city.*

iii) *Shoe shops are found in the centre✓ of the city and only along main roads.✓ Newsagents are dotted throughout the city. There are more newsagents✓ than shoe shops.*

iv) *Newsagents are found in many parts of the city as they are a convenience shop✓ and people do not want to travel far✓ to go to them. Shoe shops are found in city centres where the majority of shops are found.*

Examiner's comments

These answers are all given by the same candidate. Although the candidate is obviously a good student who has a high level of understanding of this topic s/he only scores full marks in question part iii).

i) The actual distance is 16.8 km. You would be given a margin of error in the examination and I would accept any answer between 16 and 18 km.

ii) The candidate understands the meaning of distribution but only makes two points.

iii) Three points made = 3 marks.

iv) There are only two points of explanation here. The sentence about shoe shops describes but does not *explain* as the question demands.

EXAM SPOTLIGHT

Study the OS map below.

i) Give a four figure grid reference for the Metro Centre. [1]

ii) Name the leisure facility found at grid reference 211628. [1]

iii) The area on the map marked by box A is the Central Business District of Newcastle. Use map evidence only to list three features typical of the CBD. [3]

iv) The Metro Centre, in grid square 2162, is Europe's largest out of town shopping centre. Use the OS map to describe the location of the Metro Centre. [3]

v) Explain why many people prefer to shop in this type of shopping centre rather than in the CBD. [4]

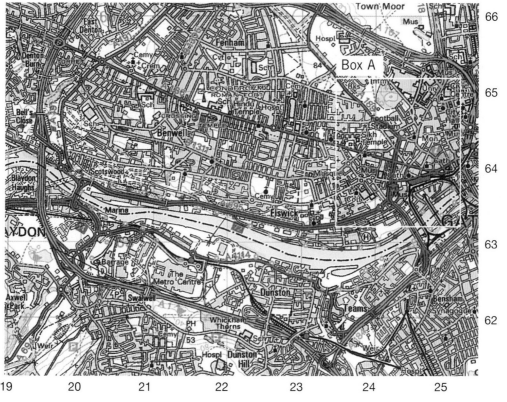

Figure 3 Part of Ordnance Survey 1:50,000 map of Newcastle upon Tyne (Landranger 88)

What are the current types of employment in Wales?

How do we classify work and employment?

The Essentials

Industry can be divided into four main types:

- **Primary** industry involves extracting resources from the sea or land and includes farming, fishing, forestry, coal mining and oil drilling. It is located where raw materials are available.
- **Secondary** industry is also known as **manufacturing** industry and makes things for people by processing raw materials or assembling components. A motor car, for example, is assembled from components made by other manufacturers.
- **Tertiary** industry provides a service and includes doctors, teachers, entertainers, shop assistants and lawyers.
- **Quaternary** industry is concerned with research, development, information and communications technology. It includes employment in medical research, finance and computer industries.

Work can also be classified as belonging to either the **public** or **private** sectors. Public sector employees include teachers, social workers and police. Private sector employees either work for themselves or for a non-government owned organisation such as Sony or Tesco. In Wales, roughly equal numbers of people work in public and private sectors.

Go Active

Play 'spot the difference' between the two pie charts in Figure 1. How many differences are there?

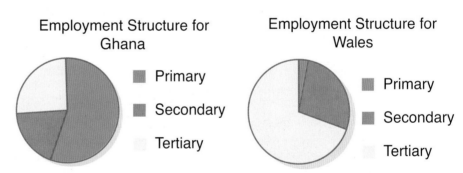

Figure 1 Employment structures in Wales and Ghana

Is there an a real pattern to this classification of work?

The Essentials

Primary industry

Less than three per cent of the population of Wales work directly in agriculture, although farming is the dominant land use for most of the country.

- Agriculture is particularly important in mid and west Wales.
- 80 per cent of farming areas are classified by the European Union as '**Less Favoured Areas**' where sheep farming or forestry are dominant.
- Dairy cattle and crops are important in lowland areas in the west, e.g. potatoes in Pembrokeshire.
- Powys is famous for beef cattle.
- North east Wales has some of the most fertile land in the country and is an important area for dairy and cereal production.

Mining and quarrying were once an important source of employment in Wales. However, today they employ less than one per cent of the population. Coal mining is now confined to a small number of privately owned mines such as the Aberpergwm drift mine in the Vale of Neath.

Slate quarrying has declined as an industry but is still a significant employer in parts of north Wales. Welsh gold, mined in near Dolgellau, is in high demand but employs few people. Some former coal mines and slate quarries have now become tourist attractions, such as Big Pit in Blaenavon.

Secondary industry

Around twenty per cent of the population of Wales are employed by the manufacturing industry. These jobs are not evenly distributed, with south Wales and the north east having the biggest concentrations.

Wales has been successful in attracting new **hi-tech manufacturing** industries. Industries such as Panasonic, Bosch and British Airways Engineering are concentrated along the **M4 corridor** in south Wales. In the north, industry is concentrated in the north east with companies such as Airbus at Broughton.

Hi-tech industry is an example of one which is **footloose**. They can easily move to new locations if the area attracts them. Bosch announced the closure of its plant near Llantrisant in 2010.

> **Footloose industry:** one which, as it is not tied to raw materials, has a free choice of location.

Tertiary industry

Tertiary industry is the biggest employer in all areas of Wales. Services are concentrated in **urban areas** and **coastal resorts**. Cities such as Cardiff, Swansea, Newport, Aberystwyth, Conway and Wrexham have important shopping centres. Cardiff is the home of the Welsh Assembly.

All large towns and cities have large numbers of people employed in local government, schools, health and transport. Coastal resorts have large numbers of people employed in the leisure industry, such as in Tenby, Barmouth, Colwyn Bay and Porthcawl.

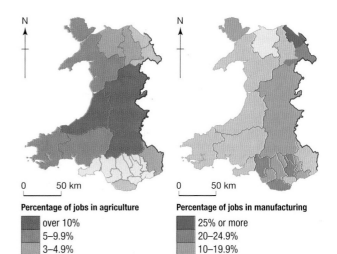

0   50 km

Percentage of jobs in agriculture

- over 10%
- 5–9.9%
- 3–4.9%
- 1–2.9%
- 0–0.9%

Percentage of jobs in manufacturing

- 25% or more
- 20–24.9%
- 10–19.9%
- 5–9.9%
- 0–4.9%

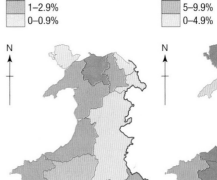
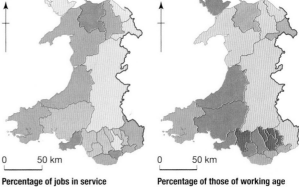

0   50 km

Percentage of jobs in service industries 2004

- more than 80%
- 76–80%
- 70–75%
- less than 70%

Percentage of those of working age who are economically inactive 2005

- more than 25%
- 22–25%
- 19–21%
- less than 19%

Go Active

Look at the maps in Figure 2 and describe the patterns that are shown. What are the differences between the top two maps (showing jobs in agriculture and manufacturing) and the bottom two maps (showing jobs in services and those that are of working age but economically inactive)?

Figure 2 Employment in Wales

What is the future of employment in Wales?

How and why are these patterns of work changing?

The Essentials

Agricultural change

- Agriculture generates only 1.2 per cent of Welsh GDP but employs 60,000 people and makes a huge contribution to the Welsh culture and landscape.
- Generally, agriculture in Wales has suffered in recent years but there are some notable exceptions. Organic farming, for example, has seen an increase in land area of about twenty per cent over the last ten years.

- Farmers in LFAs are heavily dependent on grants and subsidies from the European Union. Hill farms in particular have suffered in recent years due to competitive, cheaper imports from countries such as New Zealand, an increase in fuel and foodstuff prices and disruption from foot and mouth disease.
- Some farmers, if located in areas which have large numbers of tourists, have diversified into leisure, providing fields for caravans, bed and breakfasts and activity holidays such as paintballing and quad biking.

Case Study – Folly Farm, Pembrokeshire

Folly Farm opened in 1988 following a decision to diversify from dairy farming into the leisure industry. Today around 400,000 visitors attend the 200 acre park and, in 2005, Folly Farm won the Wales Tourist Board's 'Best Day Out in Wales' award. The farm has 60 full time employees and an additional 100 seasonal members of staff.

History
1988 – opened to the public with milking, bottle feeding and petting attractions
1989 – introduced outdoor adventure playgrounds
1992 – go-kart attraction introduced
1996 – first funfair ride installed
2001 – Follies Theatre opens and work commences on construction of the zoo
2004 – Folly Wood country park opens
2008 – Folly Farm hosts annual Industry Zoo Conference

Go Active

How else might farms diversify? List at least four ways and draw a small image next to each to help you remember it. You can research farm diversification if you are unsure.

Industrial change

The structure of the Welsh economy has been transformed over the last century.

- The valleys of South Wales were once famous for coal mining. By the 1980s the industry had virtually disappeared from the valleys. Open cast mining continues e.g. Ffos-y-Fran in Merthyr and several small drift mines exist but production is tiny in comparison to the past.
- The slate industry dominated the economy of north-west Wales during the second half of the 19th century. In 1898, a work force of 17,000 men produced half a million tons of slate. Most of the large quarries were closed by the 1970s.

- Heavy industry developed, based on natural resources and dominated the Welsh economy until the second half of the 20th century. Steelworks remain at Port Talbot but most of the heavy industry has closed or workers have been replaced due to **mechanisation**. This **de-industrialisation** has caused major social problems in parts of Wales.
- In recent years there has been a huge growth in light industries using modern technologies attracted by government incentives, an excellent infrastructure, a pleasant environment, a location inside the EU and a willing, skilled and relatively cheap workforce.

Today light manufacturing is at the heart of the Welsh economy with 1500 companies employing 75,000 people.
- Often these industries are owned by **multi-national companies** (MNCs), such as Amersham International, GE Aviation, Panasonic and Toyota.
- Wales is one of the most advanced automotive regions in the UK with over 150 companies locating here e.g. Ford and Fram filters.
- Wales has become a technologically advanced economy based on electronics, aerospace, engineering and telecommunications.
- The Welsh Assembly Government actively encourages new investment. Every new job that is created through inward investment generates many more through the **multiplier**.

Here is an example from a positive economic multiplier from www.geographyfieldwork.com

Make a copy of the outline of this flow diagram and complete it to show the effect of a firm closing down or relocating out of Wales. Use a case study on which to base your diagram.

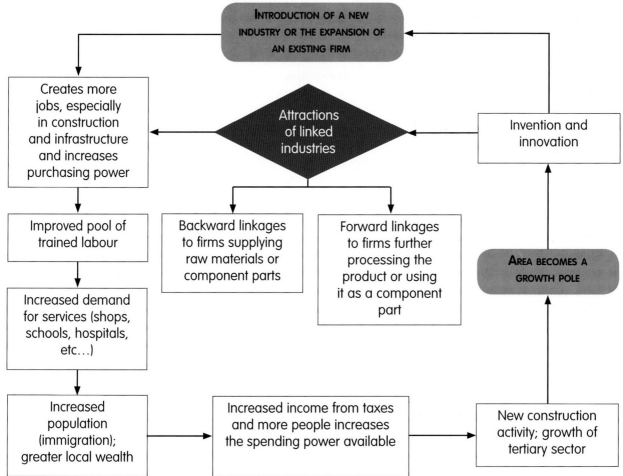

Figure 3 An example of a positive economic multiplier

Tertiary change

There has been a massive growth in service sector employment in Wales.
- Increasing living standards has led to a consequent growth in retailing, leisure and tourism, e.g. an Amazon warehouse opened in Swansea in 2008.
- There has also been growth in the percentage of the population who work in health, social work, education, transport, banking and financial services.
- The Welsh Assembly government and devolution of power has led to significant growth in government jobs.
- Companies including BT, Vodaphone, T-Mobile and NTL have bases and call centres in Wales.
- Wales also has a vibrant television and film industry with BBC Wales and S4C.

What may be the impacts of these changes?

The Essentials

Rural migration

In many of the more remote rural areas of Wales, low incomes, lack of alternative employment opportunities and entertainment facilities have led to out migration, particularly of the young and more able. This results in:

- an ageing rural population
- village shops closing and bus services stopping
- farming families becoming isolated, some suffering from loneliness
- rural schools and clinics becoming increasingly expensive and some are closing.

The problems are compounded by the sale of rural properties as holiday homes for wealthy urban dwellers who 'price out' the local people. Second home owners often spend little time in rural areas and do not support local services or add to the village community.

Deindustrialisation in the South Wales valleys

A changing employment structure has led to the following:
- A decline in jobs in mining, manufacturing and textiles and an increase in service sector jobs.
- It has led to high levels of unemployment and major social problems.
- The Welsh Assembly government has given grants and loans providing funds to promote regeneration in the valleys.
- Industrial estates, such as Treforest, were first set up to encourage industry to the area.
- New roads, such as the A470 and A4119, have improved the accessibility of the valleys.

The M4 corridor and the growth of modern industry

In south Wales much of the new industrial growth is concentrated along the M4 motorway. Access to raw materials is relatively unimportant to hi-tech industry so location, although dominated by communication considerations, can increasingly take into account the social needs of its employees. Climatic, scenic, health and entertainment factors are therefore considered and these industries locate near to places where a skilled workforce can be employed, such as the university cities of Swansea and Cardiff.

Many of the new companies that have invested in Wales are multinational companies (MNCs). Globalisation has thus benefited the Welsh economy, although some question these benefits. They point to the fact that the jobs provided tend to be low paid, 'screwdriver' jobs with the high paid, managerial and research positions remaining in the company's home country. However, the growth of new industry along the M4 corridor has led to the immigration of people, large scale house building and an increase in wealth of the population.

Go Active

Choose an MNC that has invested in Wales. Draw its logo onto a sheet of A4 paper. List three advantages and three disadvantages of this investment.

Case Study – Growth of tourism in north Wales

- North Wales has many family attractions, from zoos and farm parks, steam trains and events from cultural festivals, to agricultural shows and family fun days.
- The area offers a wide range of activities including walking, cycling, fishing, golf and water sports.
- To the west lies Snowdonia, a magnet for climbers and walkers, and on the Llyn Peninsula you will find some of the best sailing and surfing beaches in north Wales.
- The Isle of Anglesey is surrounded by 125 miles of coastline and a host of historical sites, while the north Wales borderlands are dominated by the spectacular Clwydian hills and interesting market towns on Chester's doorstep.
- The coastal resorts offer a wide range of exciting and fun filled attractions, plus a wide programme of events and evening entertainment.

Go Active

Write your Facebook or Twitter status update to describe the impact that tourism development is having on the largely rural communities of north-west Wales.

What changes are likely to take place regarding energy supply and demand in Wales?

Natural resource: something that comes from the earth that is useful to people.

How does Wales supply its current energy needs?

The Essentials

Our demand for energy has grown significantly due to increases in population, travel, entertainment and modern gadgets.

Energy needs can be sourced from renewable or non-renewable sources:

- Renewable sources include **hydroelectric**, **solar**, **wind**, **wave**, **geothermal** and **tidal power**. These energy sources will not run out.
- Non-renewable sources include **coal, oil** and **gas**. These natural resources were formed millions of years ago and are known as fossil fuels. They will eventually run out.
- **Nuclear power** is non-renewable.

Wales produces most of its energy in power stations using non-renewable fossil fuels. Only three per cent of the country's energy comes from renewable sources. Electricity is the most convenient way to transport energy around a mountainous country and many rural communities are not connected to the gas mains.

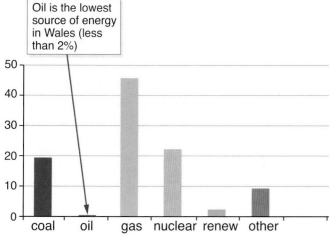

Figure 4 Sources of energy in Wales

Go Active

Describe the graph in Figure 4 by adding descriptive labels. You must give at least two descriptions of your own but you can use the modelled example to structure your descriptions.

What future changes may take place regarding energy sources and demand?

The Essentials

Using fossil fuels to generate electricity does harm to the environment and is not sustainable as they will run out in the future. Carbon dioxide produced from the burning of fossil fuels is linked to global warming. The British government signed up to the Kyoto Treaty in 1997, which is an international agreement that aims to reduce emissions of carbon dioxide in the atmosphere by 2012. The Welsh Assembly aims to make Wales a 'global showcase' for the production of renewable energy. Their target is a three per cent reduction in greenhouse gas emissions each year from 2011 onwards.

Wales has many potential renewable energy sources:

- Wind – the south west prevailing wind provides good potential for this energy source. **Wind farms** may be built on hill tops or in off shore sites.
- **Biomass** – the planting of forest and energy crops such as willow could provide this energy source.
- **Hydroelectricity** – although Wales has many rivers, none are big enough for large scale hydroelectricity schemes. Many smaller scale schemes are possible, as in Dinorwig in north Wales.
- Marine – this is another source which has good potential in Wales. The construction of a **barrage** across the river Severn could potentially provide five per cent of Britain's electricity.

Case Study – Wind power: Burbo Bank

Wind power is a controversial commodity. Although it is renewable and goes some way towards solving the problem of scarce fossil fuels for energy, some would argue that it has its drawbacks. Those who live close to wind farms claim that they are unsightly and noisy.

Burbo Bank wind farm was fully operational by the summer of 2007 and consists of 25 turbines which will produce enough energy to supply 80,000 homes. The wind conditions in this part of the Mersey estuary are perfect for harnessing wind energy. During the installation, phase noise could be heard from nearby residents as a 'thud' while piles supporting the turbines were fixed into the seabed. Some people living along the Sefton Coast have since had their view of Snowdonia blighted by the wind turbines.

Go Active

Study the chart on page 95. Choose two different sources of energy (one renewable and one non-renewable) and then complete the table below.

Energy source	What do you think are the pros and cons?		What do your parents think?		What might a local councillor think?	
	Pros	Cons	Pros	Cons	Pros	Cons

Some argue that we should be pursuing more vigorously policies to conserve energy. The British government, for example, are forcing vehicles to become more energy efficient and emit less atmospheric pollution.

Go Active

Have a look around your house. Think about your daily life and your time at school. How many ways to conserve energy can you spot? Here are two to get you started:

1 Unplugging the mobile phone charger, even when the power source is off.
2 Using an extra blanket on your bed instead of sleeping with the heating on.

Type of energy	Source	Advantages	Disadvantages
Wind	Wind turbines (modern windmills) turn wind energy into electricity.	Usually grouped together in wind farms. Potentially infinite energy supply.	Some object to them, arguing that they spoil the countryside. They depend on the wind blowing so do not produce a continuous supply of electricity.
Tidal	The movement of tides drives turbines. A tidal barrage (a type of dam) is built across estuaries, forcing water through gaps.	Could potentially generate a lot of energy. Tidal barrages can double as bridges and help prevent flooding.	Construction of barrages is very costly. Only a few estuaries are suitable. May have a negative impact on wildlife.
Hydroelectric Power (HEP)	Energy harnessed from the movement of water through rivers, lakes and dams.	Creates water reserves as well as energy supplies.	Costly to build. Can cause the flooding of surrounding communities and landscapes. Dams have major ecological impacts.
Biomass	Organic material can be burnt to provide energy, e.g. heat or electricity. An example is oilseed rape, which produces oil that can be used as a fuel in diesel engines.	It is a cheap and readily available source of energy. If replaced, biomass can be a long term, sustainable energy source.	When burnt, biomass gives off pollution including greenhouse gases. It is only a renewable resource if crops are replanted, which can take up valuable farmland.

Case Study – The Severn estuary tidal barrage

Benefits	Possible disadvantages
• It is a source of green energy and helps meet the UK's renewable energy targets.	• Existing, protected ecosystems would be heavily altered.
• The scheme has a long lifespan (over 120 years).	• Large areas of the low tide mudflats would be lost, displacing bird populations.
• It will provide flood protection for the vulnerable Severn estuary from storm surges at sea.	• It could pose a dangerous barrier to migratory fish.
• New road and/or rail transport links could be built across the barrage.	• It is likely to stimulate silting in some areas and coastal erosion in others.
• It would produce leisure-friendly water conditions behind a barrier.	• Shipping would have to navigate locks, incurring extra costs.
• It would boost the local economy – construction in the short term, tourism and infrastructure in the long term.	• All industrial discharge into the River Severn, e.g. from Avonmouth, will have to be reassessed.
• There is potential to provide up to five per cent of the UK's electricity.	• There is likely to be a negative visual impact upon the landscape.
	• A huge amount of concrete is needed which produces huge amounts of CO_2 (a greenhouse gas) when it is manufactured.

Go Active

1 Highlight the table above, using red for social, black for economic and green for environmental features.

2 Once you've done this, you could help your recall of the information by re-writing the features on a separate piece of paper. Stick to the same colours and create a table, sorting the features into their three different areas. Add pictures to help with your revision.

Answers from page 6

A	B	C	D	E	F	G	H	I	J	K
6	10	3	4	11	1	7	9	5	8	2